Governing the Uncertain

Monica Tennberg
Editor

Governing the Uncertain

Adaptation and Climate
in Russia and Finland

Editor
Dr. Monica Tennberg
Arctic Centre
University of Lapland
Pohjoisranta 4
96200 Rovaniemi
Finland

ISBN 978-94-007-3842-3   e-ISBN 978-94-007-3843-0
DOI 10.1007/978-94-007-3843-0
Springer Dordrecht Heidelberg London New York

Library of Congress Control Number: 2012933156

© Springer Science+Business Media B.V. 2012
No part of this work may be reproduced, stored in a retrieval system, or transmitted in any form or by any means, electronic, mechanical, photocopying, microfilming, recording or otherwise, without written permission from the Publisher, with the exception of any material supplied specifically for the purpose of being entered and executed on a computer system, for exclusive use by the purchaser of the work.

Printed on acid-free paper

Springer is part of Springer Science+Business Media (www.springer.com)

Preface

This book is one of the many products of a collaborative pan-Arctic research project titled "Community adaptation and vulnerability in Arctic regions" (CAVIAR). The first discussions on establishing such collaboration started in the mid-2000s as Nordic preparations for the International Polar Year (IPY 2007–2008) intensified. Credit for starting the collaborative process and seeing it through to completion goes to its parents: Grete Hovelsrud from Norway, Barry Smit from Canada, Carina Keskitalo from Sweden and Monica Tennberg from Finland. The IPY project resulted the publication in 2010 of "Community Adaptation and Vulnerability in Arctic Regions" (Hovelsrud & Smit, eds.), a volume exploring current and future sensitivities, adaptive capacities and vulnerabilities in numerous cases studies across the Arctic region.

The Finnish CAVIAR project was funded by the Academy of Finland (2007–2009). In addition to contributing chapters on flood prevention in Northern Finland and the Russian Arctic to the CAVIAR publication, the project researchers, based in the Arctic Centre of University of Lapland, extended the scope of analysis to include governance of adaptation. The central idea in the case studies in the CAVIAR research is that local adaptation cannot be understood without taking into consideration broader economic, social, and political contexts of adaptation. We wanted to take a closer look at those contexts and our combination of expertise in political science and anthropology made such an endeavour possible. This book is the outcome of the ensuing multidisciplinary discussions and an effort to combine local, regional and national perspectives on adaptation governance.

Acknowledgements

We would like to gratefully acknowledge the funding provided for this research by the Academy of Finland as well as the co-operation with our CAVIAR partners. With deep respect we also thank the residents of the Tatta District and the Hammastunturi-Kuttura *siida* for sharing their knowledge and providing tremendous support. Anna Stammler-Gossmann would like to expresses her particular gratitude to the Bopposov, Aiianitov, Rakhleev, Lopatin and Postnikov families. She also wishes to thank the administrative officials of Tatta, the local municipality of Ytyk Kyöl and the village of Chimnai for discussing flood management issues and providing her with archive materials. Terhi Vuojala-Magga would like to extend her special thanks to the members of the Hammastunturi co-operative, the members of the Magga family in the village of Kuttura, and her husband, Mauno Magga.

We also express our gratitude to Richard Foley for his help with the language checking of our contributions to the book.

Contents

Part I   Starting Points

1 **Introduction**... 3
   Monica Tennberg

2 **Adaptation as a Governance Practice**... 17
   Monica Tennberg

Part II   Russian Adaptation Governance

3 **Adaptation in Russian Climate Governance**............................. 39
   Maria Rakkolainen and Monica Tennberg

4 **The Big Water of a Small River: Flood Experiences
   and a Community Agenda for Change**....................................... 55
   Anna Stammler-Gossmann

Part III   Finnish Adaptation Governance

5 **Adaptation in Finnish Climate Governance**.............................. 85
   Monica Tennberg

6 **Adaptation of Sámi Reindeer Herding:
   EU Regulation and Climate Change**.. 101
   Terhi Vuojala-Magga

Part IV   Towards a Practice Theory of Adaptation Governance

7 **Responsibilisation for Adaptation**.. 125
   Heidi Sinevaara-Niskanen and Monica Tennberg

Index... 137

Contributors

Maria Rakkolainen Arctic Centre, University of Lapland, Rovaniemi, Finland, maria.rakkolainen@gmail.com

Heidi Sinevaara-Niskanen University of Lapland, Rovaniemi, Finland, heidi.sinevaara-niskanen@ulapland.fi

Anna Stammler-Gossmann Arctic Centre, University of Lapland, Rovaniemi, Finland, anna.stammler-gossmann@ulapland.fi

Monica Tennberg Arctic Centre, University of Lapland, Rovaniemi, Finland, monica.tennberg@ulapland.fi

Terhi Vuojala-Magga Arctic Centre, University of Lapland, Rovaniemi, Finland, vuojala-magga@suomi24.fi

Maria Rakkolainen earned a master's degree in political science with a specialisation in international relations and environmental studies from the University of Lapland (Finland) in 2008. Her thesis dealt with the multiple agency of governance and the complexity of administrative levels in the Cleaner Production programme in Northwest Russia. The research was based on analyses of relationship-building between enterprises and government in the areas of sustainable development and cleaner production.

Heidi Sinevaara-Niskanen is a doctoral candidate in the Unit for Gender Studies at the University of Lapland (Finland). She holds a master's degree in international relations and is writing her Ph.D. as a member of the Finnish Research School in Women's and Gender Studies. Her research interests combine international politics, Arctic studies and gender studies. At present she is examining the politics of sustainable development and how the social dimension of sustainable development in Arctic politics has been defined. The special focus of her work is how the equality and participation required by sustainable development are included in the social dimension of the Arctic.

Anna Stammler-Gossmann is an anthropologist from the Sustainable Development research group at the Arctic Centre of the University of Lapland (Finland). Her main interests are post-socialist socio-cultural transformation, local community adaptation and vulnerability to social and environmental changes; the concept of the North in politics, economics, and culture; and indigenous and non-indigenous identities. She carried out research from 1995 to 2002 at the Seminar für Osteuropäische Geschichte of the University of Cologne, in 2003–2004 at the Scott Polar Institute of the University of Cambridge, and in 2009 at the Center for Northeast Asian Studies of Tohoku University (Japan). Her research sites have covered different parts of Fennoscandia (Finland (Lapland) and Northern Norway (Finnmark)), Northwest Russia (the Murmansk region and the Nenets Autonomous District), and the Russian Far East (the Republic of Sakha and the Kamchatka Peninsula). Dr. Stammler-Gossmann has conducted fieldwork in the Republic of Sakha (Yakutia) since 1995.

Monica Tennberg is a research professor at the Arctic Centre of the University of Lapland (Finland). Her background is in political science with a particular interest to international relations. She defended her doctoral dissertation in 1998 on the development of relations between states and indigenous peoples during the establishment of the Arctic Council. The research drew on Michel Foucault's ideas about knowledge and power and their application to the theories of international regimes in international environmental co-operation. She has continued to study international environmental co-operation in the Arctic, most recently focusing on Arctic climate politics and the performance of international environmental co-operation in Northwest Russia. Dr. Tennberg and her research team participated in a pan-Arctic research project on community adaptation and vulnerability to climate change in Arctic regions (CAVIAR), funded by the Academy of Finland (2007–2009).

Terhi Vuojala-Magga is an anthropologist working at the Arctic Centre, University of Lapland, Finland. Her research interests include the indigenous peoples of Siberia and Northern Finland. She has carried out this work at the University of Oulu (Finland), the University of Manchester (UK), and the University of Lapland (Finland). She has specialised in the themes of climate change and reindeer husbandry, pursuing in these contexts theoretical interests in non-verbal communication and the tacit knowledge of Sámi reindeer herders. In addition to carrying out academic research, she lives and works in the Inari region and practices reindeer herding. She is finalising her doctoral dissertation on long-term environmental adaptation among the Sámi.

Abbreviations

ACIA	Arctic Climate Impact Assessment
BEAC	Barents Euro-Arctic Council
CCS	Comprehensive climate strategy for the sustainable development of the Arctic regions of Russia in the circumstances of changing climate
EMERCOM	The ministry of Russian Federation for civil defense, emergencies and elimination of consequences of natural disasters
ELY	Centre for economic development, transport and environment, Finland
EU	European Union
FICCA	Finnish climate change adaptation research
FIGARE	Finnish global change research program
FINADAPT	Assessing the adaptive capacity of the Finnish environment and society under a changing climate
FINSKEN	Developing consistent global change scenarios for Finland
GRP	Gross regional product
ICLEI	International council for local environmental initiatives, since 2003 local governments for sustainability
IPCC	Intergovernmental panel on climate change
ISTO	Climate change adaptation research programme
KGB	Soviet secret police
MAF	Ministry of forestry and agriculture
MAVERIC	Map-based assessment of vulnerability to climate change
MMM	Finnish ministry of agriculture and forestry
RHA	Reindeer herders association
RKTL	Finnish game and fisheries research institute
ROSHYDROMET	Federal service for hydrometeorology and environmental monitoring, Russia

UNFCCC	United Nations Framework Convention on climate change
SILMU	Finnish research programme on climate change
UNDP	United Nations Development Programme
VACCIA	Vulnerability assessment of ecosystem services for climate change impacts and adaptation

List of Figures

Fig. 1.1	The case study areas in the Arctic	11
Fig. 4.1	The Tatta River in Ytyk Kyöl	57
Fig. 4.2	The big water of a small river: flooding in Ytyk Kyöl in 2007	57
Fig. 4.3	Tatta District: flood prone area	61
Fig. 4.4	Regular flooding of pastures: part of a seasonal cycle in the agro-pastoralist community of Uolba	62
Fig. 4.5	Federal road in the Tatta District: the only connection with the outside world and one of the causes of the 2007 flood	65
Fig. 4.6	Emergency help: EMERCOM team during the flood in the Tatta District	67
Fig. 4.7	Failed project: water pipeline in Uolba	72
Fig. 4.8	Disaster solidarity: 'sandbag' community of Ytyk Kyöl	74
Fig. 4.9	Yard during the flood in 2007	75
Fig. 4.10	Yard after the flood in 2009	75
Fig. 6.1	Aerial view of part of the village of Kuttura	103
Fig. 6.2	EU-compliant reindeer slaughter house in Ailigas	107
Fig. 6.3	Old open-air slaughter site by the Sotajoki enclosure	108
Fig. 6.4	Reindeer leg skins for fur shoes	109
Fig. 6.5	A private reindeer meat processing facility and shop, Tundra Poro. Reindeer herder and entrepreneur Visa Valle in front of his shop	110
Fig. 6.6	Reindeer meat processed by Visa Valle	110
Fig. 6.7	Female reindeer killed by a wolf	114
Fig. 6.8	Reindeer skins on sale in front of a small cafeteria and a shop run by the Nuorgam family in the village of Kaamasmukka	118

List of Tables

Table 2.1 Two perspectives on adaptation governance 30
Table 3.1 Adaptive planning in the Murmansk region 48

Part I
Starting Points

Chapter 1
Introduction

Monica Tennberg

Abstract The chapter introduces adaptation to climate change as an issue of governance. In international climate co-operation, states have committed themselves to taking measures to ensure adaptation to climate change; the planning and implementation of such measures requires broad participation. The word "governance" refers to a multitude of actors, activities and relations between the state and other societal bodies working for the "government". The research focuses on three questions: (1) How is adaptation seen as a problem or an issue to be governed, (2) how is adaptation governed, that is, by which national, regional and local practices, and (3) what kinds of agencies do those practices enable and constrain? The analyses examine responsibility in adaptation governance at the national, regional and local levels in Finnish and Russian Arctic. The chapter introduces a research strategy that is grounded in the concept of practices and embraces political science and anthropological perspectives.

Keywords Adaptation • Governance • Responsibility • Arctic • Russia • Finland • Research questions • Research strategy • Practice theory

1.1 Adaptation in the Arctic

Adaptation to changing environmental and socio-economic conditions has been the basic requirement for human survival in the Arctic throughout the region's history. Climate change is one of the more recent challenges in this regard. Adaptation

M. Tennberg (✉)
Arctic Centre, University of Lapland, P.O. Box 122, 96101 Rovaniemi, Finland
e-mail: monica.tennberg@ulapland.fi

to climate change is fraught with uncertainty: What do Arctic peoples, communities and livelihoods need to adapt to exactly, and how? According to the current understanding, the principal threat is the pace of the expected changes vis-à-vis the ability of ecosystems, peoples and livelihoods in the Arctic to adapt to them. The impacts of climate change, both positive and negative, will not be distributed evenly across the region. Adaptation to different impacts will take place on different time scales, with some ecosystems, communities and livelihoods adapting better and more quickly than others. There may be delays in adaptation, and even maladaptations. Adaptation is not a linear and uniform process. In the Arctic, as elsewhere, the potential for adaptation, – that is, adaptive capacity, which is based on existing and future material, intellectual and organisational resources – differs among regions, communities and livelihoods. The changing climate might bring new opportunities, especially in economic terms, but taking advantage of the benefits brought by climate change might also require some adaptation. (Intergovernmental Panel on Climate Change, IPCC 1997, 2007; Arctic Climate Impact Assessment, ACIA 2004)

Adaptation is an international concern. The purpose of the United Nations Framework Convention on Climate Change (UNFCCC 1992) is to ensure that ecosystems are afforded a chance to adapt to climate change and its impacts. Food security and economic development should not be endangered because of climate change. In the Arctic, the need to adapt to climate change has emerged on the political agenda of two relevant regional forums: the Arctic Council and Barents Euro-Arctic Cooperation (BEAC). The Arctic Council is a pan-Arctic forum for discussing concerns shared by the region's states and indigenous peoples. BEAC, whose focus is the European Arctic, takes the form of collaboration between the Nordic countries, the Russian Federation and the European Commission. Where the European Arctic is concerned, the European Union (EU) is a key actor with its own plans for adaptation (Koivurova et al. 2009; Keskitalo 2010).

The rationale for governmental intervention to advance adaptation is to ensure the welfare of the population and economy and to support new opportunities brought by climate change. There are many governmental concerns related to adaptation in the Arctic; they can be described as a mix of traditional and human security concerns and risks (Yalovitz et al. 2008; Hoogensen 2009; Nuttall 2007; Lukovich 2009). For example, in some parts of the region, food security may be threatened as a result of climate change (Ford 2009a), but there are also technical and infrastructure-related threats stemming from extreme weather events and the melting of permafrost (Ford et al. 2008; McBeath 2003). Some traditional livelihoods might face a struggle in adapting to new environmental and economic conditions, and the survival of indigenous cultures facing multiple changes has been questioned (Krupnik and Jolly 2002; Ford 2009b). Additional issues in the Arctic in particular are legal and highly political concerns such as access to natural resources – oil and gas, fish, timber and minerals – and how these might be exploited in the future as ice and permafrost melt. The changes in the Arctic may require a rethinking of domestic legal practices and existing international agreements (Koivurova et al. 2009; Heinämäki 2010).

Much of adaptation takes place autonomously. Adaptation is also a purposeful action, taking place either before changes – pro-actively – or after them – re-actively. Adaptation can also be planned. Recent research and debate on governance for adaptation to climate change emphasises the importance of barriers to adaptation (Pearce et al. 2010; Hulme et al. 2007; McEvoy et al. 2010), reform of governance structures (Adger 2003; Smith et al. 2009; Martello 2008a), institutional stress management (Young 2009) and the development of enabling institutions for adaptation (Keskitalo 2010; Hovelsrud and Smit 2010; Brooks et al. 2009; Swart et al. 2009). The present study, which includes two very different cases from the Arctic – Russia and the Finnish North – with different political, economic and administrative structures for adaptation governance, undertakes to show the relevance of different nationally and locally based practices of governance for adaptation (see also Tennberg 2009; Stammler-Gossmann 2010; Vuojala-Magga 2009; Tennberg et al. 2010; Vuojala-Magga et al. 2011). It provides a contribution from the European and Russian Arctic to the recent research literature on multilevel adaptive governance and management in the Arctic, which has often focused on North America (Armitage et al. 2007; Armitage and Plummer 2010).

1.2 Responsibility in Adaptation Governance

Responsibility is a core issue in adaptation governance. With this in mind, we address the key question, "How is the responsibility to adapt and to implement adaptation measures shared between various levels and actors in adaptation governance?" Internationally, there is a "common, but differentiated" responsibility to tackle climate change and its impacts (UNFCCC 1992). Mitigation of harmful emissions that lead to climate change has been considered a more sensible solution economically than facing the uncontrolled impacts of climate change in the future (Stern 2006). However, as international climate co-operation has made only modest progress, the issue of adaptation has become more important than before. International climate co-operation is based on the assumption that climate change and its impacts are governable (see Okereke and Bulkeley 2007).

In international climate governance, adaptation is seen as an area of governmental interventions. The UNFCCC (1992) defines adaptation as a responsibility on the part of states to prevent and foresee harmful impacts of climate change, thereby making adaptation a space for governmental intervention. Responsibility is for each state and for its citizens as the state is particularly responsible for looking after its own citizens. This view also suggests that each state is primarily responsible for looking after its own citizens, not those of other countries (Hindess 2003). States have committed themselves to taking measures to ensure "adequate" adaptation to climate change.

The state has many roles to play in adaptation to climate change. The state is an adaptor itself as well as a catalyst for or regulator of adaptation to impacts of climate change for others. As service providers, governments are expected to develop their own programmes and activities to take into consideration the impacts of changing climate; this will require adaptation in governmental activities and responsibilities. As catalysts for or facilitators of adaptation, governments are to support research and the dissemination of information to societal actors on climate change impacts and adaptation options. This role also requires co-ordination between different actors and activities at various levels in order to ensure effective adaptive measures. Finally, governments can act as legislators and regulators to advance adaptation in different kinds of societal activities (Brooks et al. 2009, 5–6).

Adaptation governance is a multilevel process. Implementing current adaptation measures and making plans for future adaptation requires broad participation. From this perspective, adaptation governance is a governmental activity, but only loosely associated with the executives and bureaucracies of the formal organs of state (Dean 1999, 11). The word "governance" refers to a multitude of actors, activities and relations between the state and other societal bodies working for the "government". The current mode of global environmental governance is based on multiple stakeholders and practices of multilevel governance (Armitage et al. 2007; Keskitalo 2010).

Accordingly, other societal actors also become responsible for adaptation. For example, communities are important actors in promoting and implementing climate change strategies. It is in local communities that adaptations to impacts of climate are experienced and become concrete; these are the sites where adaptation is practiced. For example, the development of infrastructure, which includes the maintenance of roads and bridges as well as land use planning and emergency preparedness, often takes place at the community level. The national regulation that applies to these activities is operationalised at the local level (Langlais 2008; Schreurs 2008; Lundqvist and Borgstede 2008).

Because of the multilevel character of adaptation governance, the central question in our study is that of agency in adaptation, that is, "Who is the actor in adaptation and who is responsible for taking action for adaptation?" (See Jones 2004; Jagers and Duus-Otterstrom 2008; Mastrandrea et al. 2010). This research examines responsibility in adaptation governance at the national and local levels. Choosing two very different angles in theories of governance, it examines the practices shaping how the problem of climate change is understood and governed. The first approach highlights the democratic aspect of adaptation governance. Following the ideas of the American pragmatist John Dewey (1927), this "deliberative" model of adaptation governance emphasises a bottom-up approach to adaptation governance, the role of educated, responsible individuals and the importance of community deliberation in dealing with problematic situations (See also Festenstein 1997; Cochran 2002). The second approach to adaptation governance stresses the importance of power. The model originates in the thought of French historian Michel Foucault (2007), who viewed governance as a political rationality – a "governmentality" – that included ways of problematising issues, sharing power and responsibility between different actors, and producing agencies and subjectivities for governance. The approach is top-down: For Foucault, a community is a site of power relations and a technology of power (See also Agrawal 2005; Neumann and Sending 2010).

These two models view the agent of adaptation, at least on the surface, from very different point of views. The first stresses democratic participation in adaptation governance and the second views adaptation governance as practices and tactics of power in society. The two differ in how they understand adaptation as a problem and an object of governance. Despite the differences, the Deweyan and Foucauldian approaches share a critical view of the role of the state and its interactions with society and its members (Kadlec 2007; Arts et al. 2009). Their approaches to understanding the importance of self-creation of agency are similar, although by no means identical. They are both interested in how the self-creation of responsible and adaptive agents is possible (Luxon 2008; Reynolds 2004; Garrison 1998).

1.3 Research Questions

The research focuses on three questions: (1) How is adaptation seen as a problem or an issue to be governed, (2) how is adaptation governed – by which national, regional and local practices, and (3) what kind of agencies do those practices make possible, enable and constrain? First, as a rather complex environmental and economic issue, climate change reflects the multiple relations that we have to our natural surroundings as human beings. For the Arctic, climate change was initially seen in positive terms; it was only later that it became a problem, a threat or a risk, in the history of climate science. Recently, much of the current climate change debate has been framed in terms of the "opportunities" climate change will bring to the Arctic, its communities and livelihoods. It is in this light that the research first asks how concern over the climate, that is the problem of governing the human-environment relationship, has been formulated by different actors. The questions here are: What kinds of practices define our way of understanding climate change, what is the problem, how serious is it, what should be done about it, and by whom? The two case studies in the present work will investigate the differences in the national, regional and local understandings of climate change as a problem.

The second question examines the environment and human relations to it, that is, the climate, as a thing to be governed. Here, adaptation to climate change becomes a question of identifying objects (people, nature, and the relations of the former to the latter, for example, in managing the "climate" problem) to be governed and arranging the relations and responsibilities for action between the various actors involved. Our Russian and Finnish cases of national adaptation governance are intended to show the importance of traditions and practices in formulating plans and strategies for adaptation. Our local case studies, one focusing on adaptation to floods in Siberia, the other on changing conditions for reindeer herding in Finnish Lapland, explore the dynamics of local governance. Our conception of adaptation to the impacts of climate change directs our attention to the practices of governing and how they are implemented. The research asks what kind of local, national and regional practices for governing the problem have been developed or are being developed for the future. Issues of democracy and power emerge in the development of adaptation governance, prompting questions such as what kinds of activities, partnerships and networks of co-operation are established, and how, to manage, solve and govern climate change and its impacts; and who participates in these forums and how?

The third research question concerns the kinds of agencies and subjectivities that these practices of understandings and governance of climate change adaptation allow, maintain and constrain. The question of practices of agency in adaptation is an important one in the Arctic: Are the inhabitants of the Arctic "vulnerable victims of changing environmental conditions", "resilient local experts" and "indigenous advocates of sustainable development"? The agent of adaptation is at the same time an object and a subject of adapting to changing conditions. The questions are: Who adapts and how, and who are the agents in adaptation? This view suggests that adaptation is a question of democracy and power. All of us are constituted "as agents with certain capacities for action; consequently, any further control of our conduct by others or by ourselves can only be exercised in those capacities" (Owen in Neumann and Sending 2007, 692). In this sense, everyone is an actor in one way or another in adapting to the impacts of climate change, and as an outcome, an effect, of adaptation governance (See also Smith et al. 2001; Reid 2010).

1.4 Research Strategy

In the research, different knowledges – based on both political science and anthropology – explore the ways in which people, communities and governance structures have adapted and are adapting to climate change. Political science looks at international climate co-operation as a political and economic governance structure stretching from international to local levels. Adaptation to climate change is an issue of power, justice and democracy for a political scientist. Political science-based knowledge helps to understand the structural conditions for adaptation and the construction of opportunities for adaptation locally, regionally and nationally (Vanderheiden 2008; Weber 2008).

Anthropology aims at providing a perspective from the ground up, from local communities and individuals tackling and coping with various changes and challenges, of which climate change is but one (Crate and Nuttall 2009). Climate change adaptation is a question of everyday life, cultures and practices. It emphasises adaptation as lived and experienced, in the everyday life of peoples and communities. These two forms of scientific knowledge provide different accounts of adaptation practices. As knowledges, they themselves are based on social and historical practices of knowledge production, but combined they provide an important source of knowledge for developing adaptation studies beyond vulnerability and impact assessments (Mastrandrea et al. 2010).

The present research strategy embracing political science and anthropological perspectives on adaptation is based on the concept of practices. The related theory focuses attention on habitual practices of "saying and doing" adaptation. In practice theory, agents are body/minds who sustain and carry out practices. Practices are a set of routinised bodily, material, and enacted performances, but at the same time a set of mental activities, that is, background-knowledge-based activities that are patterned and meaningful (Reckwitz 2002; Schatzki et al. 2001). Practices can sometimes also be understood as skills for adaptation. Some skills are consciously taught and learned, but some are based on tacit, unarticulated, practical and commonsensical knowledge in which "the aim of a skilful performance is achieved by the observance of a set of rules which are not known as such to the person following them" (Polanyi 1956, 4 in Lynch 1997).

From our point of view, most importantly, a human subject is a relatively autonomous actor, but one whose actions are constrained and enabled by structures. Practices connect social action, in this case agency, and social structures (Berard 2005; Schatzki 1996; Turner 1994). The case studies from Russia and Finland explore the multitude of specialised "structures of collective adapting", each of which exhibits specific problems, rules of action and agencies (Franke and Roos 2005). Practices of Russian and Finnish adaptation, which exhibit very different traditions of environmental, economic and political governance, also show the differences between states and their governments in tackling the problem of adaptation and in producing agency for adaptation.

1.5 Russian and Finnish Case Studies

A case study and comparison between them is a rather typical approach to studying climate change adaptation in the Arctic (Hovelsrud and Smit 2010; Keskitalo 2008, 2010; Lange 2008). This research focuses on two very different Arctic countries and adaptation in them. The political systems in Finland and Russia are based on representative democracies, but the political structures, traditions and practices in the two countries differ considerably from one another. Since the early 1990s, Russia has gone through major political and economic changes that have resulted in a new political, administrative and economic governance structure. Adaptation to the impacts of climate change has been a focus of the Russian governmental strategy since the beginning of international climate co-operation in the early 1990s. Russia is both a major producer of fossil fuels and a greenhouse gas emitter in global terms. Export of energy resources, especially to European markets, has provided Russia with a great deal of revenue. Occupying 58% of the Arctic, it is also a country that will face major ecological changes due to climate change, including erosion and melting of permafrost. Twenty-four million people live in the Russian Arctic, a population that includes Russians as well as different indigenous groups, such as the Sami and the Nenets. (See also Keskitalo and Kulyasova 2009; Tynkkynen 2010; Rowe 2009)

Finland has been a member state of the European Union since 1995. Therefore, its climate politics follow and are influenced by EU decisions and policies on climate change (see Ellison 2010). The Finnish economy is very energy intensive. Finland is a minor producer of greenhouse gases in global terms, although the country's level of emissions is high in relation to its population. Its Arctic region, Finnish Lapland, covers almost one third of the territory of the country; it is home to a population of some 180,000 people, of whom 4,000 are Sami. Recently, Finland has drawn attention to the vulnerability of the people and nature in Northern Finland, in its 2006 and 2010 national communications to the UNCCC. The country is at the forefront of international efforts in its preparedness to adaptation. Its national adaptation strategy was formulated in 2005 and updated in 2009. (Kerkkänen 2010; Teräväinen 2010; Jauhola 2010)

Our political science-based research examines the national plans for adaptation in Russia and Finland in detail. These two countries and their efforts to adapt to the impacts of climate show the diversity of conditions and opportunities for adaptation nationally, regionally and locally. The analyses of the national adaptation strategies examine policy statements, vulnerability assessments and national reports. The research focuses on the way adaptation to climate change has been identified as a problem, what kinds of activities are being undertaken and planned for the future to support adaptation, and who is responsible, and how, for seeing to it that adaptation takes place. One of the researchers in the Finnish case study had an opportunity for participant observation in 2010–2011 as a member of the steering group for a project developing a regional climate strategy for Lapland.

The two anthropological case studies (see Fig. 1.1) explore adaptation governance at the local level. The Russian local case study examines flooding as an adaptation concern in two Siberia communities along the Tatta River in the northern part of the Sakha Yakutia Republic. During the Soviet era, a dense water management network was developed in the region of Tatta River. Until recently the inhabitants along river saw themselves living in a stable environment, but in the last few years exceptional, heavy floods have challenged this view. The anthropological case study investigates how people and communities in the region adapt to flooding as a process of negotiation. The study is based on the data collected during intensive field work conducted in 2008 and 2009 in the settlements of Ytyk Kyöl and Uolba along the river.

The Finnish anthropological case study comes from the village of Kuttura in the Hammastunturi-Kuttura reindeer herding collective in Finnish Lapland. The research focuses on adaptation by local reindeer herders in the Kuttura siida, a Sami self-governance system, to changing conditions of reindeer work in terms of environmental, economic and legal requirements. The Kuttura *siida* comprises five households and a total of some 20 members, including children. The local case study explores adaptation governance as a possibility to negotiate and act within a group of reindeer herders, as well as action in these circumstances in response to the changes taking place within reindeer husbandry.

The present case studies from two different Arctic countries – Finland and Russia – illuminate human agency in adaptation under different historical, environmental, geographical, cultural, political and administrative conditions. The approach in this research emphasises that human agents are simultaneously situated in different social problematics and in different fields of practices. The

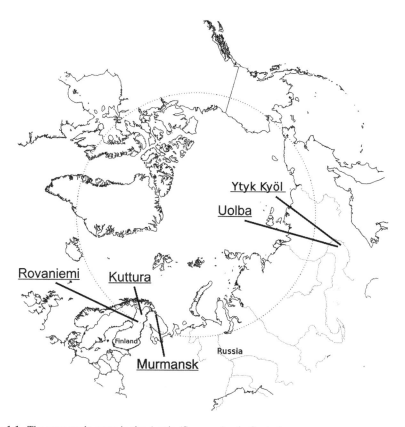

Fig. 1.1 The case study areas in the Arctic (Source: Arctic Centre)

case studies explore the dynamics of adaptation from two angles and levels, one top-down and the other bottom-up. The difference between top-down and bottom-up perspectives may be understood as a "standing controversy", but as Haila and Dyke (2006, 29) suggest, such a controversy is "false"; both perspectives are needed and necessary to explore the process of constructing responsibility in adaptation governance.

In the context of international climate co-operation, there has been a particular subject who has become authorised to speak on behalf of the environment: the Arctic indigenous peoples (Krupnik and Jolly 2002; Martello 2008b; Vuojala-Magga 2009). In this research, the focus is not the adaptation of indigenous peoples to climate change. Our point of view is that everyone living in the Arctic, indigenous or non-indigenous, is an object of adaptation policies and strategies and governed by decisions made at the local, regional, national, and even international level. On balance, it makes sense to study the adaptation of Arctic peoples, livelihoods and communities in general.

1.6 The Structure of the Book

The book is divided into four parts: starting points, Russian adaptation governance, Finnish adaptation governance and conclusions. In part I, Chap. 2 by Monica Tennberg presents two theoretical approaches to the study of adaptation governance in the context of climate change. The first approach – deliberative adaptation governance – is based on what may be considered the classical liberal political tradition. It draws on ideas inspired by the pragmatist thought of John Dewey, an American philosopher and educator who emphasised individual growth, education and community deliberation. From this perspective, adaptation to climate change and the development of adaptation governance is an opportunity for social learning and reform. The other perspective, more critical of the classical liberal tradition, is based on the ideas of French philosopher and social historian Michel Foucault. It views adaptation as a question of power tactics and struggles between the state and other actors, including various levels of governance. From this perspective, development of adaptation governance results in adaptive agencies. These two extremes highlight the multitude of governance options for climate change adaptation when it comes to formulating the problem and developing governance practices and agencies for adaptation.

In part II, the focus is on Russian adaptation governance. Chapter 3 by Maria Rakkolainen and Monica Tennberg, which deals with the Russian national adaptation strategy, focuses on the development of Russian adaptation governance practices. The chapter describes and analyses how climate change and its impacts are understood as a problem in Russia. It also discusses the political rationales underlying Russian climate politics and the role of adaptation in it with regard to the Arctic region. The chapter also presents some information about regional and local efforts to tackle the question of adaptation in the Murmansk region.

Chapter 4 by anthropologist Anna Stammler-Gossmann studies local adaptation to flooding as it is reflected in the interaction of local conditions and forces at the regional and national levels in Russia, specifically in communities along the Tatta River in Siberia. Environmental risks, such as floods, are framed through national strategies of governing, such as "emergency" and "risk". However, for local Tatta residents, flooding is one among many other problems and threats in everyday life. It is not a "risk" in the Western understanding dominant in international climate cooperation. For local Tatta residents, noticing the links between flooding, the creation of wealth and risks may be seen as challenging tasks.

In part III, the focus is on Finnish adaptation governance. Chapter 5 by Monica Tennberg presents and discusses the Finnish national adaptation strategy and governance practices. The possible impacts of climate change in Finland have been studied extensively since the 1990s. Moreover, adaptation planning has developed quickly and spread widely from national to local levels. The chapter describes and discusses the development of Finland's understanding of the problem of climate change and the need to adapt to it. Concern over the Arctic is a novelty in the Finnish national debate. The research also describes regional and local efforts to tackle the question of adaptation in Finnish Lapland.

Chapter 6 by Terhi Vuojala-Magga builds up a debate on Sami reindeer husbandry in the context of climate change and EU politics. The anthropological case study focuses on the production and sale of reindeer meat and policies regarding predators in the Inari region of northern Finnish Lapland. The concepts of agency and governance are defined from the local perspective, in which the actor is the individual reindeer herder and the concept of governance is linked to the traditional Sami kinship system, or *siida*. The research examines the strengths and dynamics of a *siida* and its individual members where climate change is concerned within the jungle of EU regulations and constraints.

In the concluding part IV, Chap. 7 by Heidi Sinevaara-Niskanen and Monica Tennberg brings together and analyses the foregoing chapters to formulate an understanding of responsibility in climate change adaptation based on practice theory. The responsibility for adaptation is shared between international, regional, and local adaptation governance practice: in our understanding it is formed in a process which we call "responsibilisation". Responsibilisation is a practice of government "to shape, normalize and instrumentalize the conduct, thought, decisions and aspirations of others in order to achieve the objectives they [authorities] consider desirable" (Miller and Rose 1990, 7). The focal question is the kind of agency for adaptation that governance constructs. Who is responsible for taking action? Political agency is thus involved in the formulation and choice of who are regarded as capable of governing themselves and who are not (Neumann and Sending 2007, 699). The opportunities for adaptation are closely linked to political and economic governance structures. The adaptive agents are enabled, constrained and created by our existing political, administrative and economic structures. Our approach highlights the questions of responsibility and power in adaptation through practices of adaptation.

References

ACIA (2004). Arctic climate impact assessment. http://amap.no/acia/. Retrieved 17 May 2010.
Adger, W. N. (2003). Social capital, collective action and adaptation to climate change. *Economic Geography, 79*(4), 387–404.
Agrawal, A. (2005). *Environmentality. Technologies of government and the making of subjects*. Durham/London: Duke University Press.
Armitage, D., & Plummer, R. (Eds.). (2010). *Adaptive capacity and environmental governance*. Berlin: Springer.
Armitage, D., Berkes, F., & Doubleday, N. (Eds.). (2007). *Adaptive co-management. Collaboration, learning and multi-level governance*. Vancouver: UBC Press.
Arts, B., Lagendijk, A., & Houtumn, H. (Eds.). (2009). *The disoriented state: Shifts in governmentality, territoriality and governance*. Berlin: Springer.
Berard, T. J. (2005). Rethinking practices and structures. *Philosophy of the Social Sciences, 35*, 196–230.
Brooks, M., Gagnon-Lebrun, F., & Sauvé, C. (2009). *Prioritizing climate change risks and actions on adaptation: A review of selected institutions, tools and approaches*. Ottawa: Policy Research Initiative.
Cochran, M. (2002). A democratic critique of cosmopolitan democracy: Pragmatism from the bottom-up. *European Journal of International Relations, 8*(4), 517–548.

Crate, S. A., & Nuttall, M. (Eds.). (2009). *Anthropology and climate change: From encounters to actions*. Walnut Creek/Oxford: Left Coast Press/Berg.

Dean, M. (1999). *Governmentality: Power and rule in modern society*. London: Sage.

Dewey, J. (1927). The public and its problems. In J. Boydston (Ed.), *John Dewey. The later works 1925–1953* (Vol. 2). Carbondale: Southern Illinois University.

Ellison, D. (2010). Addressing adaptation in EU policy framework. In C. Keskitalo (Ed.), *Developing adaptation policy and practice in Europe: Multi-level governance of climate change* (pp. 39–96). Berlin: Springer.

Festenstein, M. (1997). *Pragmatism and political theory*. Cambridge: Polity Press.

Ford, J. (2009a). Vulnerability of Inuit food systems to food insecurity as a consequence of climate change: A case study from igloolik, Nunavut. *Regional Environmental Change, 9*(2), 83–100.

Ford, J. (2009b). Dangerous climate change and the importance of adaptation for the Arctic's Inuit population. *Environmental Research Letters, 4*(2), 1–9.

Ford, J. D., Pearce, T., Gilligan, J., Smit, B., & Oakes, J. (2008). Climate change and hazards associated with ice use in northern Canada. *Arctic, Antarctic, and Alpine Research, 40*(4), 647–659.

Foucault, M. (2007). *Security, territory and population. Lectures at the collège de France 1977–1978*. Houndmills/New York: Palgrave-McMillan.

Franke, U., & Roos, U. (2005, August). *From collective actor to structure of collective acting. The meaning of human beings for the study of international relations*. Paper presented at the first world international studies conference, Istanbul and the third conference for the European consortium for political research, Budapest.

Garrison, J. (1998). Foucault, Dewey and self-creation. *Educational Philosophy and Theory, 30*(2), 111–134.

Haila, Y., & Dyke, C. (2006). Introduction. What to say about nature's 'speech'. In Y. Haila & C. Dyke (Eds.), *How nature speaks. The dynamics of human ecological condition* (pp. 1–47). Durham: Duke University Press.

Heinämäki, L. (2010). *The right to be a part of nature: Indigenous peoples and the environment* (Acta universitatis lapponiensis 180). Rovaniemi: Lapland University Press.

Hindess, B. (2003). Responsibility for others in the modern state system. *Journal of Sociology, 39*(1), 23–30.

Hoogensen, G. (2009). Security at the poles: The Arctic and Antarctic. In H. G. Brauch, U. Spring, J. Grin, C. Mesjasz, P. Kameri-Mbote, N. Behera, B. Chourou, & H. Krummenacher (Eds.), *Facing global environmental change. Environmental, human, energy, food, health and water security concepts. Part VIII* (pp. 951–960). Berlin: Springer.

Hovelsrud, G., & Smit, B. (Eds.). (2010). *Community adaptation and vulnerability in Arctic regions*. Berlin: Springer.

Hulme, M., Adger, N. W., Desai, S., Goulden, M., Lorenzoni, I., Nelson, D., Naess, L-O., Wolf, J., & Wreford, A. (2007). Limits and barriers to adaptation: Four propositions. Tyndall Centre briefing note 20. http://www.tyndall.ac.uk/sites/default/files/bn20.pdf. Retrieved 11 May 2011.

Intergovernmental Panel on Climate change, IPCC. (1997). *The regional impacts of climate change: An assessment of vulnerability*. Cambridge: Cambridge University Press.

Intergovernmental Panel on Climate change, IPCC. (2007). Polar regions (Arctic and Antarctic). In M. L. Parry, O. F. Canziani, J. P. Palutikof, P. J. van der Linden, & C. E. Hanson (Eds.), *Climate change 2007: Impacts, adaptation and vulnerability. Contribution of working group II to the fourth assessment report of the intergovernmental panel on climate change* (pp. 653–685). Cambridge: Cambridge University Press.

Jagers, S. C., & Duus-Otterstrom, G. (2008). Dual climate change responsibility: On moral divergences between mitigation and adaptation. *Environmental Politics, 17*(4), 576–591.

Jauhola, S. (2010). Mainstreaming climate change adaptation: The case of multilevel governance in Finland. In C. Keskitalo (Ed.), *Developing adaptation policy and practice in Europe: Multi-level governance of climate change* (pp. 149–188). Berlin: Springer.

Jones, R. N. (2004). Incorporating agency into climate change risk assessments. An editorial comment. *Climatic Change, 67*, 13–36.

Kadlec, A. (2007). *Dewey's critical pragmatism*. Lanham: Lexington Books.

Kerkkänen, A. (2010). *Ilmastonmuutoksen hallinnan politiikka. Kansainvälisen ilmastokysymyksen haltuunotto suomessa* (Acta universitatis tamperensis 1549). Tampere: Tampere University Press.

Keskitalo, E. C. H. (2008). *Climate change and globalization in the Arctic: An integrated approach to vulnerability assessment*. London: Earthscan.

Keskitalo, C. (Ed.). (2010). *Developing adaptation policy and practice in Europe: Multi-level governance of climate change*. Berlin: Springer.

Keskitalo, E. C., & Kulyasova, A. (2009). Local adaptation to climate change in fishing villages and forest settlements in northwest Russia. In S. Nystén-Haarala (Ed.), *The changing governance of renewable natural resources in northwest Russia* (pp. 227–243). Aldershot: Ashgate.

Koivurova, T., Keskitalo, C., & Bankes, N. (Eds.). (2009). *Climate governance in the Arctic*. Berlin: Springer.

Krupnik, I., & Jolly, D. (2002). *The Earth is faster now: Indigenous observations of Arctic environment change*. Fairbanks: ARCUS.

Lange, M. (2008). Assessing climate change impacts in the European north. *Climatic Change, 87*, 7–34.

Langlais, R. (2008). *Climate change emergencies and European municipalities: Guidelines for adaptation and response*. Stockholm: Nordregio.

Lukovich, J. V. (2009). Addressing human security in the Arctic in the context of climate change through science and technology. *Mitigation and Adaptation Strategies for Global Change, 14*(8), 697–710.

Lundqvist, L. J., & Borgstede, C. (2008). Whose responsibility? Swedish local decision-makers and the scale of climate change abatement. *Urban Affairs Review, 43*(3), 299–324.

Luxon, N. (2008). Ethics and subjectivity. Practices of self-governance in the late lectures of Michel Foucault. *Political Theory, 36*(2), 377–402.

Lynch, M. (1997). Theorizing practice. *Human Studies, 20*, 335–344.

Martello, M. L. (2008a). Vulnerability analysis and environmental governance. In P. Dauvergne (Ed.), *Handbook of global environmental politics* (pp. 417–431). Cheltemham: Elgar.

Martello, M. L. (2008b). Arctic indigenous peoples as representations and representatives of climate change. *Social Studies of Science, 38*(3), 351–376.

Mastrandrea, M. D., Heller, N. E., Root, T. L., & Schneider, S. H. (2010). Bridging the gap: Linking climate-impacts research with adaptation planning and management. *Climatic Change, 100*, 87–101.

McBeath, J. (2003). Institutional responses to climate change: The case of the Alaska transportation system. *Mitigation and Adaptation Strategies for Global Change, 8*(1), 3–28.

McEvoy, D., Matczak, P., Banaszak, I., & Chorynski, A. (2010). Framing adaptation to climate-related extreme events. *Mitigation and Adaptation Strategies for Global Change*. doi:10.1007/s11027-010-09233-2.

Miller, P., & Rose, N. (1990). Governing economic life. *Economy and Society, 19*, 1–31.

Neumann, I. B., & Sending, O. J. (2007). The international as governmentality. *Millennium, 35*(3), 677–701.

Neumann, I. B., & Sending, O. J. (2010). *Governing the global polity. Practice, mentality, rationality*. Ann Arbor: University of Michigan.

Nuttall, M. (2007). An environment at risk: Arctic indigenous peoples, local livelihoods and climate change. In J. B. Ørbæk, R. Kallenborn, I. Tombre, E. N. Hegseth, S. Falk-Petersen, & A. H. Hoel (Eds.), *Arctic alpine ecosystems and people in a changing environment* (pp. 19–35). Berlin: Springer.

Okereke, C., & Bulkeley, H. (2007). Conceptualizing climate change governance beyond the international regime: A review of four theoretical approaches. *Tyndall centre working paper 112*. http://www.tyndall.uea.ac.uk/book/export/html/297. Retrieved 17 May 2010.

Pearce, T., Ford, J. D., Duerden, F., Smit, B., Andrachuck, M., Berrang-Ford, L., & Smith, T. (2010). Advancing adaptation planning for climate change in the Inuvialuit settlement region (ISR): A review and critique. *Regional Environmental Change*. doi:10.1007/s10113-010-0126-4.

Reckwitz, A. (2002). The status of the "material" in theories of culture: From "social structure" to "artefacts". *Journal for the Theory of Social Behavior, 32*(2), 195–217.

Reid, J. (2010, May). *The debased and politically disastrous subject of resilience*. Key note speech at the ARKTIS graduate school seminar. Rovaniemi: University of Lapland.

Reynolds, J. M. (2004). "Pragmatic humanism" in Foucault's later work. *Canadian Journal of Political Science, 37*(4), 951–977.

Rowe, E. W. (2009). Who is to blame? Agency, causality, responsibility and the role of experts in Russian framings of global climate change. *Europe-Asia Studies, 61*(4), 593–620.

Schatzki, T. R. (1996). *Social practices. A Wittgensteinian approach to human activity and the social*. Cambridge: Cambridge University Press.

Schatzki, T. R., Knorr-Cetina, K., & von Savigny, E. (2001). *The practice turn in contemporary theory*. London: Routledge.

Schreurs, M. A. (2008). From the bottom-up: Local and sub-national climate change politics. *The Journal of Environmental Development, 17*(4), 343–355.

Smith, J., Lavender, B., Smit, B., & Burton, I. (2001). Climate change and adaptation policy. *ISUMA, Winter*, 75–81.

Smith, J. B., Vogel, J. M., & Cromwell, J. E., III. (2009). An architecture for government action on adaptation. An editorial comment. *Climatic Change, 95*, 53–61.

Stammler-Gossmann, A. (2010). 'Translating' Vulnerability at the community level: Case study from the Russian north. In G. K. Hovelsrud & B. Smit (Eds.), *Community adaptation and vulnerability in Arctic regions* (pp. 131–162). Berlin: Springer.

Stern, N. (2006). Stern review on the economics of climate change. http://www.webcitation.org/5nCeyEYJr. Retrieved 18 May 2010.

Swart, R., et al (2009). *Europe adapts to climate change. Comparing national adaptation strategies*. PEER report 1. Vammala: Saastamalan kirjapaino.

Tennberg, M. (2009). Is adaptation governable? Regional and national approaches to climate change. In T. Koivurova, C. Keskitalo, & N. Bankes (Eds.), *Climate change governance in the Arctic* (pp. 289–302). Berlin: Springer.

Tennberg, M., Vuojala-Magga, T., & Turunen, M. (2010). The Ivalo river and its people. There have always been floods – What is different now? In G. K. Hovelsrud & B. Smit (Eds.), *Community adaptation and vulnerability in Arctic regions* (pp. 221–238). Berlin: Springer.

Teräväinen, T. (2010). Political opportunities and storylines in Finnish climate policy negotiations. *Environmental Politics, 19*(2), 196–216.

Turner, S. (1994). *The social theory of practice. Tradition, tacit knowledge and presuppositions*. Cambridge: Polity Press.

Tynkkynen, N. (2010). A great ecological power in global climate policy? Framing climate change as a policy problem in Russian public discussion. *Environmental Politics, 19*(2), 179–195.

UNFCCC (1992). United nation framework convention on climate change. http://unfccc.int/resource/docs/convkp/conveng.pdf. Retrieved 17 May 2010.

Vanderheiden, S. (Ed.). (2008). *Political theory and global climate change*. Cambridge: MIT.

Vuojala-Magga, T. (2009). Simple things, complicated skills: Archaeology, practical skills and climatic change from the perspective of anthropology. In T. Äikäs (Ed.), *Mattut maddagat. The roots of saami ethnicities, societies and spaces/places* (pp. 164–173). Saastamala: Vammalan Kirjapaino Oy.

Vuojala-Magga, T., Turunen, M., Ryyppö, T., & Tennberg, M. (2011). Resonance strategies of Sámi reindeer herders in northernmost Finland during climatically extreme years. *Arctic, 64*(2), 227–241.

Weber, E. P. (2008). Facing and managing climate change: Assumptions, science and governance response. *Political Science, 60*(1), 133–149.

Yalovitz, K. S., Collins, J. F., & Virginia, R. A. (2008). *The Arctic climate change and security policy*. Final conference proceedings. http://carnegieendowment.org/files/arctic_climate_change.pdf. Retrieved 17 May 2010.

Young, O. R. (2009). Institutional dynamics: Resilience and vulnerability in environmental and resource Regimes. *Global Environmental Change, 20*, 378–385.

Chapter 2
Adaptation as a Governance Practice

Monica Tennberg

Abstract In general terms, governance is the exercise of political, scientific, economic and administrative power to manage societies and their development. The concept of adaptation governance captures the multitude of issues, activities and actors engaged in adaptation to climate change. Adaptation to the impacts of climate change is a problem for governance, but one that is understood in different ways. Accordingly, the chapter develops two theoretical approaches to the study of adaptation governance, one drawing on the work of John Dewey, the other on the ideas of Michel Foucault. Adaptation governance does not appear from a political, economic and societal vacuum, but is closely related to existing political and administrative habits, customs and practices. Governance enhances or changes relations of responsibility. As a result of "responsibilisation" – increasing responsiveness in adaptation governance – practices of governance and power are developed, agencies are constructed and responsibilities for adaptation shared.

Keywords Adaptation • Governance • Responsibilisation • John Dewey • Michel Foucault • Governmentality

2.1 Adaptation and Responsibility

Adaptation to changing environmental, economic and social conditions has been a research interest in the Arctic for a long time. A rather new research topic in adaptation studies is governance for adaptation to consequences of human-induced climate change.

M. Tennberg (✉)
Arctic Centre, University of Lapland, P.O. Box 122, 96101 Rovaniemi, Finland
e-mail: monica.tennberg@ulapland.fi

According to the IPCC (2007, 720), adaptation to climate change includes adjustments in practices, processes or structures to take account of changing environmental conditions. Much of the current research focuses on adaptive capacity which is presented as an approach to assess and support "the ability of potential of a system to respond successfully to climate variability and change, includes adjustments in behaviour and in resources and technologies" (ibid., 727). This understanding of adaptation has come to dominate the debate on international climate co-operation and science (Orlowe 2009; see also Janssen and Ostrom 2006).

In addition to adaptive capacity, current research focuses on various kinds of "adaptivities": adaptive learning (Keeney and McDaniels 2001), adaptive co-management (Armitage and Plummer 2010), adaptive policies (Swanson et al. 2010; Walker et al. 2001) and adaptive institutions (Glaas et al. 2010; Keskitalo 2010; Koivurova et al. 2009). The word "adaptive" has a positive connotation; it refers to interactive, open and learning processes of adaptation. Adaptive capacity is seen as an analytical tool for societal planning in general rather than for planning particular adaptation measures to match specific climate risks and opportunities (Smit and Pilifosova 2003, 12). Adaptive capacity is important for "transition management", which combines a long-term vision of sustainable development with short-term experimental learning to tackle problems, probe options and find pathways to realise the vision for sustainable development (Voβ et al. 2009, 277).

There are, however, a number of problems related to the concept of adaptive capacity and closely related concepts as they are currently used. These ideas are action-oriented, but are limited in their technical understandings of the issues and processes to be covered and often emphasise the calculative logic of adaptation (Orlowe 2009, 135–136). In contrast to adaptive capacity, adaptation is critically described as reactive, rigid and path-dependent (Staber and Sÿdow 2002, 411). Path-dependency is a feature in institutional development that constrains future choices: "Path-dependence is a way to narrow conceptually the choice set and link decision-making through time. It is not a story of inevitability in which the past neatly predicts the future" (North 1990, 98–99). Path-dependency in climate politics is often understood in terms of barriers to adaptation (Garrelts and Lange 2011).

The strength of path-dependency is that it makes understandable how current policies have accumulated over time and how previous choices in decision-making restrict current and future options for policy-makers. However, for our purposes, path-dependency is not enough. It is a temporal organising principle that looks more backwards than forwards and is unable to capture different scales and levels of governance. Most importantly, it explains the bounded nature of institutional rationality and human choices by economical thinking (Kay 2005, 554).

In general terms, governance is the exercise of political, scientific, economic and administrative power to manage societies and their development. In our view, the concept of "adaptation governance" captures the multitude of issues, activities and actors engaged in adaptation. It does not claim that governance is "adaptive" but it embraces adaptation in one way or another. Adaptation governance does not appear from a political, economic and societal vacuum, but is closely related to existing habits, customs and practices. It is not merely a question of economically based choices and rationalities leading to path-dependencies or path-change, nor is it based on various processes of managerialism to support and increase "adaptivities" in society.

In our view, governance enhances or changes relations of responsibility (Pellizzoni 2004, 542). Responsibility can be understood in terms of *care* (responding to pre-set normative and cultural beliefs concerning the human-nature relationship and how one should behave according to them), *liability* (responding to things that have happened, for example in connection with natural hazards and accidents), *accountability* (justifications for choices and actions taken and to be taken, for example in climate politics and decision-making), or as *responsiveness* to a present situation without pre-set beliefs or after-the-fact evaluations and explanations of the situation, as a process to cope with uncertainties (ibid., 549).

In adaptation governance, following Pellizzoni's idea, responsiveness can be understood as a means for interpreting ongoing changes and a tool for evaluating the development of adaptation governance and its ability to cope with future environmental challenges. It allows us to investigate various kinds of temporalities – past, present and future – and their various mixes that emerge in adaptation governance. As a result of "responsibilisation" – increasing responsiveness in adaptation governance – practices of governance and power are developed, agencies are constructed and responsibilities for adaptation shared. It also makes it possible to search for an answer to the question of how responsibilities are made for adaptation (O'Neill et al. 2008, 214). The development of adaptation governance depends on and reflects the existing governmental structures, and political practices, including the role of citizens as political actors. The practice theory approach allows the study of adaptation governance as a process and as a result, as the emergence of something that is not known yet. It enables us to study not only the habitual and customary aspects of adaptation, but also new practices in the making.

2.2 Problem of Adaptation

Adaptation is often also used as an alternative to and in connection with other concepts, such as resilience and vulnerability. Resilience refers to the capacity of ecosystems and populations to recover from stress or change. In the Arctic context, resilience has been used by natural science based climate change research in particular (Chapin et al. 2004, 2006; Berkes and Jolly 2001; Forbes 2008). Vulnerability has also been used with reference to the Arctic, often in research on climate change in the social sciences (Turner et al. 2003; ACIA 2004; Sydneysmith et al. 2010). The concept of vulnerability has been applied in studying the attributes of persons or groups that enable them to cope with the impact of disturbances, particularly in the case of natural hazards (Adger 2006; Smit and Wandel 2006).

Adaptation to the impacts of climate change is a problem for governance, but Deweyan and Foucauldian approaches understand the nature of the problem in different ways. The Deweyan approach focuses on "problematic situations" and finding solutions for them, whereas the Foucauldian views "problems" as problematisations, that is, particular constructions to be governed. American philosopher and educator John Dewey (1859–1952) saw humans as adaptive to their environments and capable of learning. In the Deweyan approach, adaptation draws on flexible human capabilities to learn to adapt in different conditions (McDonald 2004, 78). Human experience is based on recurrent problematicity as "a normal state of affairs" (Turnbull 2008). For Dewey, the world was precarious and perilous, a circumstance which necessitated human adaptation. Adaptation for Dewey was a problem-solving process with many stages. The process begins with a problematic situation, a situation where habitual human responses to the environment are inadequate for the continuation of activities. A problem is a state of doubt, hesitation, perplexity, a mental difficulty. From this perspective, the impacts of climate change are something disturbing and unwanted, and concrete elements in the everyday life of people, livelihoods and communities which will require some public problem solving. The aim of the problem solving is to determine what the problem is, what the alternatives for action are and what their consequences will be. (McDonald 2004; Festenstein 1997)

Dewey saw knowledge as arising from an active adaptation of the human organism to its environment. Much knowledge is commonsensical, but problematic situations challenge such common sense. For Dewey, developing knowledge and acting in the world were all part of the same process of learning and field of experience. The public sphere originated from the attempts to analyse and resolve problems, especially those that were caused by indirect consequences of human activities. Problematic situations involve conflicts, but they are part of problem-solving efforts and provide an opportunity for social growth. In particular, our time is characterised as one of "radical uncertainty" (Pellizzoni 2003, 328), which derives from the complexities of defining problems and of the required responses to them. For example, climate change and adaptation to it lack a simple description, an account of the connections among related facts, as well as a shared vision of meanings and the action required. Debate about climate change is accompanied by distrust of expert knowledge and conflicting roles of experts in politics. As a result, the current public sphere is fragmented by incommensurable claims of knowledge (see Pellizzoni 2003).

Dewey was accused of social engineering, especially since he promoted scientific knowledge and rationality in problem solving. However, science, according to Dewey, is not be used to manipulate the public, but to "educate and to empower" (Kaufman-Osborn 1992, 258). For Dewey, science, knowledge and democracy were closely connected. He saw a number of ways to expand our understanding of what knowledge is, who produces it and how by producing "communities of inquiry". From a Deweyan perspective, such communities are an attempt to extend the problem-solving procedures and co-operation used in science into the political sphere of action. Dewey was concerned about the participation of individuals in such arrangements. His approach stresses that individuals are motivated to seek the common good to the extent that they see their activity as a socially recognised contribution to the co-operative process (Pellizzoni 2003, 348–349).

The French philosopher and social historian Michel Foucault (1926–1984) provides another view on practices of problematisation. In the Foucauldian approach, different forms of knowledge and products of knowledge compete, construct and challenge adaptation and how it is understood. Knowledge about climate change impacts, vulnerabilities and adaptation forms the basis for governmental actions and interventions. This kind of knowledge is used in adaptation governance in two senses: *connaissance* and *savoir*. Scientific, instrumental knowledge about impacts and their social and economic consequences (*connaissance*) is important for the constitutive knowledge needed for adaptation governance (*savoir*) (Foucault 1972, 185; Tennberg 2000, 52). This latter aspect of knowledge as *savoir*, or knowledge about human beings and their activities, serves the uses of power. The Foucauldian idea is that "a living being" is "inseparable from an ever-changing grid of knowledge through which we understand life itself and its meaning" (Rose 2007, 42).

International climate governance requires a specific kind of knowledge for governing which consists of different kinds of measurements as well as the costs and benefits of adaptation governance; Oels (2005) calls this "economic" knowledge. Through this hegemonic economic knowledge, dangers, threats and risks become technologies for governing populations and their welfare (Turnheim and Tezcan 2009). As Hulme (2008) suggests, our understanding of climate change as a threat, a catastrophe and a risk is linked to neoliberal globalism and the emergence of the global risk society (see also Carvalho 2005). Liverman (2008) identifies three different narrative frames for understanding climate change: "dangerous climate change", "common but differentiated responsibility for climate change" and, finally, "climate change as an investment opportunity".

However, the costs and benefits of adaptation are not easily measured, quantified and calculated. For governmental purposes, knowledge is constructed and used to make some dangers, risks and threats related to climate change manageable (Higgins 2001). Social problems, such as adaptation to climate change, are constructed as objects of governance which Foucault terms "problematisations". A problematisation is not "simply a collection of the relevant elements per se, but also the system of relations ... established between these elements to be governed" (Dean 1999, 28). From this perspective, climate change adaptation can be considered a problematisation consisting of scientific information, local experiences, administrative plans and strategies, and international obligations. The debates surrounding climate change and adaptation by various actors frame issues as "problems" around which governmental concern then revolves, seeking to rectify the failings and to cure the ills. In the Foucauldian sense, "it is around these difficulties and failings that programmes of government have been elaborated" (ibid., 29).

These two perspectives on adaptation as a problem, one more bottom-up and the other top-down, relate to practices of questioning. The questions to study are: How are adaptation-related problems defined, and by whom? What is the problem in adaptation? What remains un-problematised? How serious are the problems caused by climate change considered to be? What needs to be done about them? And by whom? Who is responsible and who is not? Who knows what the problems and solutions are? Whose knowledge is and will be used? Practices of questioning highlight the two views of climate change: a problematic situation experienced locally by individuals and their communities or a particular governmental problematisation concerning the welfare of population. Responding to climate change and taking responsibility for action require the use of different kinds of knowledges and contributions by different knowledge providers (Pettenger 2007; Glover 2006). In the Arctic, different kinds of knowledges and the role of multiple knowledge providers has been recognised as an important element in dealing with environmental issues, including climate change (Tennberg 2000; Nilsson 2007; Shadian and Tennberg 2009).

2.3 Governance of Adaptation

A problematic situation or a problematisation requires action, a response. International climate governance is based on the assumption that adaptation to climate change and its impacts is governable (see Okereke and Bulkeley 2007; Weber 2008). Our two approaches to adaptation governance provide different views regarding the governmental response to climate change impacts: John Dewey viewed a problematic situation as part of a planning society (in contrast to a planned society; Healey 2009; Shields 2003) and Michel Foucault considers that situation as a particular problematisation for governmental intervention (Fischler 2000; Colebatch 2002). Practices of governance are central to our perspective.

One of our theoretical thinkers, John Dewey, viewed human action as habitual: "Habits constantly adapt the environment, including the social environment, to our needs, or accommodate our needs to the demands of the environment". Habits refer to dispositions and skills of individuals and for their resulting actions (Cohen 2007, 779). At the collective level, Dewey used the term "customs" for habitual human actions. Following this logic, framing governance as grounded in habits and customs, it becomes much more of an historical being (Cohen 2007, 777). Dewey called for the development of flexible habits that allow us to adapt to and direct the course of a constantly changing world (Hildreth 2009, 795). From this perspective adaptation governance is an opportunity for social learning, public participation and democracy through community deliberations to deal with problematic situations caused by climate change and its impacts. A current example of such experimentation is the idea of co-management (Armitage et al. 2007, 3) as a sharing of power and responsibility between the state and other resources users for adaptive governance.

In Foucauldian thinking, adaptation governance is a question of practices of power between the state and other societal actors. The centre of these power relations lies in the historical transformations of statehood and related understandings and practices of governmental responsibilities (Dean 1999; Lemke 2007; Jessop 2007; Fang 2009). From this perspective, adaptation governance is a governmental programme, with its particular logic, strengths and weaknesses. This perspective emphasises the mentality of governing adaptation – the process of forming a political rationality for adaptation by the state but by other societal actors as well.

For adaptation governance, the state has to adapt its own activities to changing conditions, but it also has to support others through knowledge, regulations and financing in their efforts to adapt (Brooks et al. 2009; Finan and Nelson 2009). This leads to new practices of governance and new responsibilities, split among societal actors in multilevel climate "governmentalities" that include citizens as both objects and subjects of governing (Agrawal 2005; Glover 2006; Neumann and Sending 2010).

The central question in these two approaches is the governance response in respect to democracy, as well as the power of different actors and levels of action. The Deweyan approach values the technical knowledge of experts, but calls for participatory democracy to improve "the methods and conditions of debate, discussion and persuasion". To Dewey, "this is *the* problem of the public" (Dewey 1927, 208). The focal issue is different ways of organising responsibilities for adaptation governance. Recognising a problematic situation as a problem creates a relationship between the state and the society.

Dewey saw democratic politics as residing within a community (Cochran 2002, 530). He defined democracy as more than a form of government: it is primarily a mode of associated living, of conjoint, communicated experience that secures the liberation of power. This approach thus highlights the role of the community as a site of democratic deliberation in adaptation governance. The community is the realm in which the collective will and opinion are formed. The community members are understood to be equals and as engaging in interaction that comprises discourse, inquiry and action. The Deweyan approach stresses that societal reforms are best accomplished where individuals participate actively and regularly in public affairs so that society as a whole may take advantage of their diverse experience and knowledge (Cochran 2002, 529).

Another element in Dewey's thinking on democracy is how democracy is organised politically. Dewey's starting point in thinking about political organisation is human acts and their consequences for others. A public sphere is constituted when indirect consequences are recognised and there is an effort to regulate them. The public is organised and made effective by means of representatives who regulate the joint actions of individuals and groups. An association then develops into a political organisation and something which may be governed comes into being. The public becomes a political state. The state, as a social arrangement, is a practical forum for conducting experimentation: It creates a public sphere which allows individuals to communicate with one another (Dewey 1927, 35).

In this sense, the political institutionalisation of publics involves developing organisational rules for both representation and critical discussion. Publics develop "traits of state" when they develop strong organisational and decision-making capacities and seek to make their concerns authoritative. This view challenges any strict division between states and civil society. The central issue in the current nation-state-based climate politics is to create institutional opportunities for different types of political actors who claim to represent different and overlapping constituencies to participate in solving problematic situations (Bray 2009, 716). From a Deweyan perspective, climate change adaptation is a problematic situation that could lead to the development of a public sphere. For a democratic citizen, the central issue is not the power to influence decisions, but the opportunity to participate in the related deliberation and access to that deliberation (Bohman 1999, 503). The development of international climate governance through the work of the UNFCCC and the IPCC has advanced the creation of such an international public sphere with public participation. There are a number of different ways an international public sphere can form internationally: through networks (Goodin and Dryzek 2006), stakeholder democracy (Bäckstrand 2006) and international public spheres (Cochran 2002).

The main precondition for democracy is the social division of labour: Each member of society, the public sphere, should see him- or herself as sharing a common consciousness of responsibility and co-operation. Responsible action emerges from publics constituted by persons who recognise a need for social co-operation in resolving common problematic situations (Bray 2009, 692). The Deweyan understanding of democracy and power directs our attention to the lived experiences of individuals engaged in problem-solving, prompting further study of the multiple forces and interactions that constitute the conditions for social action (Hildreth 2009, 800). For Dewey, habit is "the seat of human power to act in the world". Power is above all "the effective means of operations; ability or capacity to execute…" (Dewey 1916, 246). Power is positive: it is power to do work and accomplish things in the world. Power is also what makes the ability to execute plans – to realise aims – possible. Individual capacities are important in Dewey's understanding of power, but they are relative to a context of experience. For Dewey, power is also "the sum of conditions available for bringing the desirable end into existence" (ibid., 246). Power operates in multiple ways and is enacted through experiences. Human capacities for action are "not merely individual possessions, but a part of a complex transactional field of individual and social forces that constitute particular situations" (Hildreth 2009, 799).

The other view about governmental response, the Foucauldian, focuses on the rationality of government, that is, how it creates a way or system of thinking about the nature of the practice of government (who can govern, what governing is, and what or who is governed). Governmentality makes some form of activity thinkable and practicable to both its practitioners and those upon whom it is practiced (Foucault 1991; Dean 1999). "Government" here does not refer here to the state, but the relations between different governmental organs, officials and strategies that define the limits of a state. In these actions, the question is how and to what extent the state is or is not articulated in the activity of government (Dillon 1995). Dean defines the aim of governmentality studies as being "to understand how different locales are constituted as authoritative and powerful, how different agents are assembled with specific power and how different domains are constituted as governable and administrable" (Dean 1999, 29).

From this perspective, the current international climate governance constructs climate change as an issue of securing economic growth and population management. As an international regime, the UNFCCC and the Kyoto Protocol produce global climate change as an entity governable by advanced liberal government. The Kyoto mechanisms have transformed the atmosphere from a sewer into a regulated market in a process of commodification. Following economic logic, international climate co-operation aims at providing climate stability and makes economic growth the entity to be protected from excessive climate protection costs. The current climate governance seeks to create markets that will allow actors at all levels to make responsible, that is, cost-aware, choices regarding climate change (Oels 2005; see also Liverman 2008; Glover 1999).

In this perspective, a community is a particular technology, or a practice, of government. For Foucault, this is "to govern without governing society, to govern through regulated choices made by discrete and autonomous actors in the context of particular commitments to families and communities" (Rose 1996, 328). Here, responsibility is no longer understood as a relationship with the state, but as one of obligation to those for whom the individual cares most: his or her family, neighbourhood, workplace and, ultimately, community. And, as a result, "behaviour is increasingly governed through the realm of ethics, whereby individuals are ethically obliged to ac for the benefit of their group – to become masters of their won collective destinies by becoming ethical citizens of their community" (Summerville et al. 2008, 697). In the case of climate change, issues of adaptation are made local: localities, communities and regions are transformed into governmental sites for adaptation strategies and plans (Agrawal 2005; Schofield 2002). But Foucault is also a "thin" communitarian. He does not speak of community in a liberal sense, as an integrated, consensual, state-centred conception of political life, but rather as one that is decentred, open and dynamic and an alternative to current politics. Ideally, as Olssen reads Foucault, the state provides the basic freedoms for individuals and communities in order to constitute a decentralised democracy, a network of co-ordinative institutions and self-creative agencies (See also Olssen 2002, 507–508).

The Deweyan ideal would be deliberative adaptation governance based on broad participation of stakeholders, co-operation between different knowledge providers and a mix of knowledges, such as adaptive co-management (Armitage and Plummer 2010). From this perspective, climate change is a problem leading to the formation of a multilevel public sphere in which stakeholders can participate and, in some cases, also influence adaptation governance in different ways. The Foucauldian approach is a governmental process in which technologies of agency, including individuals and communities, are created and used to ensure economic growth and population management. Governmental concerns are transformed into justifications for interventions and actions by an assemblage of different societal actors – not only the state, but also actors affiliated and co-operating with different state organs and officials. As a global governmentality, the UNFCCC is based on a scientifically defined apparatus of global climate change and market-based practices of governance. The community is a site for the implementation of such governance practices and responsibilities. (Agrawal 2005, 6–7; Lovecraft 2008) As practices of governance, the Deweyan and Foucauldian approaches highlight the practices of governing climate change adaptation. The questions are: What kinds of relations between democracy and power do communities involve and what are their powers to deal with the impacts of climate change? How do they use their powers in adapting to climate change? What effects does community-based climate governance have in terms of power and democracy? What kind of institutional arrangements secure democracy and public participation?

2.4 Agency in Adaptation

The central question in adaptation governance is individual autonomy, democracy and freedom in adaptation. The critical issue for liberal theory is how to govern a population of free individuals such that its members accept the legitimacy of the work of government. Government takes the freedom and agency of those who are governed as both an end and a means for governing. (Hindess 2005; Ockwell et al. 2009) Governing through freedom while tackling global environmental problems such as climate change may require new kinds of citizenship, including new freedoms and responsibilities. Andrew Dobson (2003) suggests "post-cosmopolitan ecological citizenship", which should rest more on obligations than rights, on private responsibilities than public duties, on virtues of care and compassion and ideas of non-territoriality. Dobson's idea focuses on the private and voluntary sphere of action at the expense of citizenship as a political practice within the state or as part of public collective action (Wilson 2006).

In Deweyan terms, participation of citizens in democracy is a key issue. For him, democracy is a way of living together and a community is an opportunity for self-realisation. Responsible action emerges from publics constituted by persons who recognise a need for social co-operation in resolving common problematic situations (Bray 2009, 714). Direct citizen participation (Roberts 2004) is often presented as an ideal for democracy, but one that places high demands on individuals as political actors and participants. For Dewey, the greatest challenge for responsible citizens is not to accumulate knowledge that would make them policy experts; nor should responsible citizens endeavour to develop political skills that would transform them into political entrepreneurs. The basic requirement for responsible citizens is to have the ability to link seemingly isolated personal and local situations to their public and more global consequences. This requires that democratic citizens be willing to look outside their own surroundings for discussions and suggestions that might improve their understanding of their lives and its conditions (Kosnowski 2005, 672). Improving communication is a key issue, but the problem is that "there are too many publics and too much of public concern for our existing resources to cope with" (Kadlec 2007, 95–96).

Ideally, Deweyan "climate citizens" become able to see their ecological responsibility in making wise choices in their everyday life and consumption patterns. They are able to use knowledge relating to climate change and to translate it into action in their lives with an awareness of the implications their actions and choices have locally, regionally and globally. Climate citizens are able to deliberate and participate in local, regional and global discussions about climate change as well as measures to tackle the problem and to adapt to changing conditions. They are able to participate in and contribute to political action in various forums and forms of climate action. From the pragmatist point of view, the challenge is to use knowledge and education as means for capacity-building and empowering citizens to become climate citizens. For Dewey, human beings as free but responsible citizens, educated and engaged, will actively take part in such experiments. Responsible people understand the consequences and broader implications of their actions (See also Hewitt 2007; Cohen 2007).

Webber (2001) asks, "Why can't we be Deweyan citizens?" Her explanation is the limits of forgetting, that is, the search for consistency, repetition and meaning in our everyday experiences. A typical feature of the current politics is scattered, mobile and manifold publics in which individual citizens "find themselves deprived of the ability to locate either their present experience or their plans for the future within a coherent context". As a result, citizens retreat into a form of apathetic privatism which undermines their capacity to appreciate and act upon the interest that they share with others (Kaufman-Osborn 1984, 1146–1147). Dewey's theory addresses the private pre-conditions of the fragmentation of the public sphere.

As an outcome, in the "advanced welfare liberalism" of international climate governance, everyone is made responsible for climate change and action through processes of "responsibilization" (Summerville et al. 2008; Pellizzoni 2003). As a result, we might have "adaptive climate citizens", new environmental subjects "that emerge as a result of involvement in struggles over resources and in relation to the new institutions and changing calculations of self-interest and notions of the self" (Agrawal 2005, 16); or as "consumerist climate citizens" that model themselves as knowledgeable, calculating and responsible individuals who "increase their competitiveness in a constant drive for self-optimisation in keeping with economic principles" (Oels 2005). Such technologies of agency that make us climate citizens of various sorts seek to deploy our possibilities of agency as a tactic or mode of governing (Potter 2009; Ockwell et al. 2009.) These technologies engage all of us as active and free citizens, as informed and responsible consumers, as members of self-managing communities and organisations, as political actors in democratising social movements, and as agents capable of taking control of our own risks (Dean 1999, 167–168). But Tobias notes, discussing Amartya Sen's ideas on capabilities in connection with Foucault's thinking, "persons capable of forging their own ethical-political project should be left to do so", but it is also important to recognise, as Sen does, that "not all persons, at all points in their life, may be so capable" (Tobias 2005, 83).

It is these practices of agency and self-creation that both Dewey and Foucault were interested in (Luxon 2008; Reynolds 2004; Garrison 1998). The questions are: What kinds of agents does adaptation require and construct? Who are the agents? Who are not? Are they fragile, vulnerable or resilient? Are they engaged, interested and educated? Are they knowledgeable, rational and responsible? According to Dewey, democracy as an ethical ideal calls upon men and women to build communities in which the necessary opportunities and resources are available for every individual to realise fully his or her particular capacities and powers through participation in political, social and cultural life. Individuality or the capacity for autonomous choice is realisable only in a certain sort of community. Such a community offers an adequate range of options from which an individual can make a choice as well as a democratic character that fosters intelligent choice (Festenstein 1997, 22–23).

Governmentality encompasses the way in which human conduct is governed through various practices of individual self-government, the government of others and the government of the state (Allen 2002; Cruikshank 1999). Self-government and politics are easily at odds with each other: The aim of stability and harmony in ethical self-governance is challenged by politics of dissent, and contestation makes participation in politics difficult and challenging (Luxon 2008, 397). Power as a strategic situation produces "actors", enables and constrains them, shapes their fields of opportunities and limits their freedoms. Resistance to strategic power includes reacting to external demands or fulfilling pre-set expectations. For Foucault, subjects are situated within practices that in a sense constitute and organise human activity. Practices of power create the space within which we become available to ourselves and to each other as subjects of action and knowledge and as objects of knowledge and control. However, Foucault also suggests that resistance can be tactical, as action taking place within the existing practices of governance. The question for tactical resistance is, "How can the present constellation of governing practices be employed such that they enable us to forge new kinds of agency…?" (Thompson 2003, 126). From this perspective, responsibilisation of societal actors to climate change adaptation is a tactic of power by the state but also a tactic of resistance for those who are being adapted and are adapting. Ethical self-creation as adaptive subjects and objects is "a new way of refusing to be what we are". Foucault saw ethico-political practices of the self and developing relationships with others as a form of resistance to tactics of state power (Reynolds 2004, 955).

2.5 Adaptation as Governance Challenge

The concept of adaptation is historically connected to the development of modern biology and the theory of evolution. In the nineteenth century, "the classical view of a timeless continuity of nature was replaced by a concept of life in which species were understood as discontinuous entities shaped by the evolutionary influence of the environment" (Rutherford 2000, 115). Modern biology sees life as organisms dependent and linked to their external surroundings in terms of exchanging resources. Foucault's idea of biopower suggests that life has become an issue of power in different forms: allowing, controlling, and administering life. It has become an issue of distributing the living, biological subjects as efficiently as possible within the social and economic fields (ibid., 115). Nowadays, adaptation is increasingly understood as a challenge for a social organisation: It has become a question of the "ability to survive in the face of its unalterable features and the capacity to cope with uncertainty and unpredictable variations" (a quote by Parsons 1964, in Staber and Sÿdow 2002, 410).

Our two approaches, Deweyan and Foucauldian, focus on adaptation, life and evolution but in different ways. Dewey suggested that "humans are in and of nature: they live in a natural environment and their own organic and bodily processes are natural". Humans must adjust to their environment or to prevailing conditions as much as any other living creatures. Flexibility is a basic requirement for human survival and the survival of human organisations. (McDonald 2004, 69–71)

In Foucault's opinion, the environment is made up of "the conjunction of a series of events produced by these individuals, populations, and groups, and quasi natural events which occur around them" (Foucault 2007, 21). For Foucault, the evolution of the human race is a question of bio-power. Through practices of governance, the body (both individually and collectively) has become both the raw material of power and, at the same time, that which produces and transforms itself as a living being. As a result, bio-power disciplines the body of the individual to increase its utility and manageability. Bio-power also supervises the collective body – the species body – through interventions and regulatory controls (Rutherford 2000; Rose 2001).

Adaptation to climate change is not an isolated issue but one that is linked to more general debates about human adaptation and governance. Practice theory provides an approach to study agents and their structures of action from the perspective of doing and saying adaptation. The question is what kinds of practices of questioning are developed to define and understand the problem of climate change. Following Dewey's idea of problematic situations and Foucault's problematisations, the focal questions are: How are problems defined? How serious they are? How should they be solved and by whom? Who knows what the problems and solutions are? Whose knowledge is important? These questions define the responsibilities for creating problems and solving them.

As practices of governance, these two different approaches highlight the challenges of governing climate change adaptation when broaching the issue of the sites of governance, in particular the role of communities. The questions are: What kinds of relations of democracy and power do communities consist of? How do communities use their power role in adapting to climate change? What effects on power and democracy does community-based climate governance have? Finally, the practices of agency constitute the adaptive agents in one form or another. The questions to ask are: What kinds of agents does adaptation require and create? Who are the agents? Who are not?

The Deweyan and Foucauldian approaches to adaptation governance constitute bottom-up and top-down approaches, respectively. The Deweyan bottom-up idea of adaptation governance entails the development of a(n international) public sphere, based on the indirect consequences of human action and recognition of those consequences. Deliberation requires knowledge that comes from many sources and is combined for the use of decision-making in a democratic community. Free, educated and interested democratic citizens are able and willing to participate in public deliberation. In the Deweyan approach, adaptation is a case of social experimentation: Social projects are to be tested by their human consequences and how they fulfil practical social needs. What works is what benefits people: what benefits

Table 2.1 Two perspectives on adaptation governance

	Deweyan deliberation	Foucauldian governmentality
Practices of questioning	Impacts of climate change as disturbance, annoyance, a problematic situation; knowledge as experience about the problematic situation	Climate change as a governmental problematisation; knowledge as *connaissance* and *savoir*
Practices of governance	Climate change co-operation as an international public sphere; community as a site for democratic deliberation; state and society intertwined	International climate governmentality; community as a site of power relations; tactics of state power
Practices of agency	Free, but responsible, educated and engaged citizens; power relations as lived experiences and social practices; critical inquiry as a participatory practice	Free, but responsible, knowledgeable and calculable subjectivities; power relations as lived, critical self-governance

people can often be determined by thoughtful experimentation with new and untried social institutions and arrangements.

Foucault suggests that governmentality is an incomplete, ever-failing exercise in top-down governing. Its very establishment entails the beginnings of its fall and failure. Governmentalities are based on political rationalities that define what is doable and possible – and doable and possible now – by whom and with what practices. Knowledge is produced and used for governmental needs. In governmentality, technologies of agency create and maintain special subjectivities. Responsible, knowledgeable and engaged climate citizens take on responsibilities in their own lives and adjust them to the impacts of climate change. In governmentality, communities are sites of governing and of the re-organisation of power relations (Agrawal 2005; Dean 1999). The following table summarises the two perspectives discussed here (see Table 2.1).

In the following chapters, the research goes on to focus on two cases of adaptation governance, one Russian and one Finnish. Both Deweyan and Foucauldian ideas of adaptation governance are to be found in these cases. In terms of international climate governance practices, both Deweyan and Foucauldian approaches seem critical of the prospects of global institution building for dealing with climate change and its consequences (Held 2009; see also Beck 2008; Kadlec 2006). The crucial issue here is the role of the state. Both Dewey and Foucault are critical of the state and its actions. The key question for Foucault is the tactics of power in sharing responsibility for adaptation governance. For him, criticism is a historical investigation of statehood and its power practices that have led us to constitute ourselves and to recognise ourselves as subjects of what we are thinking, doing and saying (Reynolds 2004, 959). For Dewey, the state by its very nature is ever something to be scrutinised, investigated, and assessed. The Deweyan critical approach to the state consists of two questions, the first being the extent to which the public is organised and the second the degree to which state officials serve the public interest (Festenstein 1997, 85; Kadlec 2006). The case studies will explore the role and practices of the Russian and Finnish states in developing and implementing adaptation governance and the local experiences of those governance practices.

References

ACIA (2004). Arctic climate impact assessment. http://amap.no/acia/. Retrieved 17 May 2010.
Adger, W. N. (2006). Vulnerability. *Global Environmental Change, 16*(3), 268–281.
Agrawal, A. (2005). *Environmentality. Technologies of government and the making of subjects.* Durham/London: Duke University Press.
Allen, A. (2002). Power, subjectivity and agency: Between Arendt and Foucault. *International Journal of Philosophical Studies, 10*(2), 131–149.
Armitage, D., & Plummer, R. (Eds.). (2010). *Adaptive capacity and environmental governance.* Berlin: Springer.
Armitage, D., Berkes, F., & Doubleday, N. (Eds.). (2007). *Adaptive co-management. Collaboration, learning and multi-level governance.* Vancouver: UBC Press.
Bäckstrand, K. (2006). Democratizing global governance? Stakeholder democracy after the world summit on sustainable development. *European Journal of International Relations, 12*(4), 467–498.
Beck, U. (2008). Reframing power in the globalized world. *Organization Studies, 29*, 793–804.
Berkes, F., & Jolly, D. (2001). Adapting to climate change: Social-ecological resilience in a Canadian Western Arctic community. *Conservation Ecology, 5*(2), 18. http://www.consecol.org/vol5/iss2/art18. Retrieved 26 May 2010.
Bohman, J. (1999). International regimes and democratic governance: Political equality and influence in global institutions. *International Affairs, 75*(3), 499–513.
Bray, D. (2009). Pragmatic cosmopolitanism: A Deweyan approach to democracy beyond the nation-state. *Millennium, 37*(3), 683–719.
Brooks, M., Gagnon-Lebrun, F., & Sauvé, C. (2009). *Prioritizing climate change risks and actions on adaptation: A review of selected institutions, tools and approaches.* Ottawa: Policy Research Initiative.
Carvalho, A. (2005). Governmentality of climate change and the public sphere. http://repositorium.sdum.uminho.pt/handle/1822/3070. Retrieved 17 May 2010.
Chapin, F. S., Peterson, G., Berkes, F., Gallaghan, T. V., Angelstam, P., Apps, M., Beler, C., Bergeron, Y., Crépin, A.-S., Danell, K., Elmqvist, T., Folke, C., Forbes, B., Fresco, N., Juday, G., Niemelä, J., Shvidenko, A., & Whiteman, G. (2004). Resilience and vulnerability of northern regions to social and environmental change. *Ambio, 33*(6), 344–349.
Chapin, F. S., III, Hoel, M., Carpenter, S. R., Lubchenco, J., Walker, B., Callaghan, T. V., Folke, C., Levin, S., Mäler, K.-G., Nilsson, C., Barrett, S., Berkes, F., Crépin, A.-S., Danell, K., Rosswall, T., Starrett, D., Xepapadeas, T., & Zimov, S. A. (2006). Building resilience and adaptation to manage Arctic change. *Ambio, 35*(4), 198–202.
Cochran, M. (2002). A democratic critique of cosmopolitan democracy: Pragmatism from the bottom-up. *European Journal of International Relations, 8*(4), 517–548.
Cohen, M. D. (2007). Reading Dewey: Reflections on the study of routine. *Organization Studies, 28*, 773–786.
Colebatch, H. K. (2002). Government and governmentality: Using multiple approaches to the analysis of government. *Australian Journal of Political Science, 37*(3), 417–435.
Cruikshank, B. (1999). *The will to empower. Democratic citizens and other subjects.* Ithaca: Cornell University Press.
Dean, M. (1999). *Governmentality. Power and rule in modern society.* London: Sage.

Dewey, J. (1916/1967–87). Force and coercion. In J. Boydston (Ed.), *John Dewey. The middle works of John Dewey, 1899–1924* (Vol. 10). Carbondale: Southern Illinois University Press.

Dewey, J. (1927/1981–1991). The public and its problems. In J. Boydston (Ed.), *John Dewey. The later works of John Dewey, 1925–1953* (Vol. 2). Carbondale: Southern Illinois University.

Dillon, M. (1995). Sovereignty and governmentality: From the problematics of the new world order to the ethical problematics of the world order. *Alternatives, 20*, 323–368.

Dobson, A. (2003). *Citizenship and the environment*. Oxford: Oxford University Press.

Fang, S.-Z. (2009). Governing environment: Governmentality in global climate politics. *The International Journal of the Humanities, 7*(10), 17–34.

Festenstein, M. (1997). *Pragmatism and political theory*. Cambridge: Polity Press.

Finan, T. J., & Nelson, D. R. (2009). Decentralized planning and climate adaptation: Toward transparent governance. In N. W. Adger, I. Lorenzoni, & K. O'Brien (Eds.), *Adapting to climate change. Thresholds, values, governance* (pp. 335–349). Cambridge: Cambridge University Press.

Fischler, R. (2000). Communicative planning theory: A Foucauldian assessment. *Journal of Planning Education and Research, 19*, 358–368.

Forbes, B. (2008). Equity, vulnerability and vesilience in social-ecological systems: A Contemporary example from the Russian Arctic. *Equity and the Environment, 15*, 203–236.

Foucault, M. (1972). *Archaeology of knowledge*. London: Tavistock Publications.

Foucault, M. (1991). Governmentality. In G. Burchell, C. Gordon, & P. Miller (Eds.), *The Foucault effect. Studies in governmentality* (pp. 87–104). Chicago: The University of Chicago Press. With two lectures by and an interview with Michel Foucault.

Foucault, M. (2007). *Security, territory and population. Lectures at the Collège de France 1977–1978*. Houndmills/New York: Palgrave-McMillan.

Garrelts, H., & Lange, H. (2011). Path dependencies and path change in complex fields of action: Climate adaptation policies in Germany in the realm of flood risk management. *Ambio, 40*, 200–209.

Garrison, J. (1998). Foucault, Dewey, and self-creation. *Educational Philosophy and Theory, 30*(2), 111–134.

Glaas, E., Jonsson, A., Hjerpe, M., & Andersson-Skold, Y. (2010). Managing climate change vulnerabilities: Formal institutions and knowledge use as determinants of adaptive capacity at the local level in Sweden. *Local Environment, 15*(6), 525–539.

Glover, L. (1999). Atmosphere for sale: Inventing commercial climate change. *Bulletin of Science Technology Society, 19*, 501–510.

Glover, L. (2006). *Postmodern climate change*. London: Routledge.

Goodin, R. E., & Dryzek, J. S. (2006). Deliberative impacts: The macro-political uptake of mini-publics. *Politics & Society, 34*(2), 219–244.

Healey, P. (2009). The pragmatic tradition in planning thought. *Journal of Planning Education and Research, 28*, 277–292.

Held, D. (2009). Restructuring global governance: Cosmopolitanism, democracy and the global order. *Millennium, 37*(3), 535–547.

Hewitt, R. (2007). *Dewey and power. Renewing democratic faith*. Rotterdam: Sense.

Higgins, V. (2001). Calculating climate: Advanced liberalism and the governing of risk in Australian drought management. *Journal of Sociology, 37*(3), 299–316.

Hildreth, R. W. (2009). Reconstructing Dewey on power. *Political Theory, 37*(6), 780–807.

Hindess, B. (2005). Politics of government: Michel Foucault's analysis of political reason. *Alternatives, 30*, 389–413.

Hulme, M. (2008). The conquering of climate discourses of fear and their dissolution. *The Geographical Journal, 174*(1), 5–16.

IPCC. (2007). Assessment of adaptation practices, options, constraints and capacity. In M. L. Parry, O. F. Canziani, J. P. Palutikof, P. J. van der Linden, & C. E. Hanson (Eds.), *Climate change 2007: Impacts, adaptation and vulnerability. Contribution of working group II to the fourth assessment report of the intergovernmental panel on climate change* (pp. 717–743). Cambridge: Cambridge University Press.

Janssen, M. A., & Ostrom, E. (2006). Resilience, vulnerability, and adaptation: A cross-cutting theme of the international human dimensions programme on global environmental change. *Global Environmental Change, 16*(3), 237–239.
Jessop, B. (2007). From micro-powers to governmentality: Foucault's work on statehood, state formation and state power. *Political Geography, 26*, 34–40.
Kadlec, A. (2006). Reconstructing Dewey: The philosophy of critical pragmatism. *Polity, 38*(4), 519–542.
Kadlec, A. (2007). *Dewey's critical pragmatism*. Lanham: Rowman & Littlefield.
Kaufman-Osborn, T. V. (1984). John Dewey and the liberal science of community. *The Journal of Politics, 46*(4), 1142–1165.
Kaufman-Osborn, T. V. (1992). Pragmatism, policy science and the state. In J. E. Tiles (Ed.), *John Dewey. Critical assessments* (pp. 244–266). London: Routledge.
Kay, A. (2005). A critique of the use of path dependency in policy studies. *Public Administration, 83*(3), 443–571.
Keeney, R. L., & McDaniels, T. L. (2001). A framework to guide thinking and analysis regarding climate change policies. *Risk Analysis: An International Journal, 21*(6), 989–1000.
Keskitalo, C. (2010). *Developing adaptation policy and practice in Europe: Multi-level governance of climate change*. Berlin: Springer.
Koivurova, T., Keskitalo, C., & Bankes, N. (Eds.). (2009). *Climate governance in the Arctic*. Berlin: Springer.
Kosnowski, J. (2005). Artful discussion: John Dewey's classroom as a model of deliberative association. *Political Theory, 33*(5), 654–677.
Lemke, T. (2007). An indigestible meal? Foucault, governmentality and state theory. *Distinktion, 15*, 43–65. http://www.thomaslemkeweb.de/publikationen/IndigestibleMealfinal5.pdf. Retrieved 17 May 2010.
Liverman, D. M. (2008). Conventions of climate change: Constructions of danger and the dispossession of the atmosphere. *Journal of Historical Geography*. doi:10.1016/j.jhg.2008.08.008.
Lovecraft, A. (2008). Climate change and Arctic cases: A normative exploration of social-ecological system analysis. In S. Vanderheiden (Ed.), *Political theory and climate change* (pp. 91–120). Massachusetts: MIT.
Luxon, N. (2008). Ethics and subjectivity. Practices of self-governance in the late lectures of Michel Foucault. *Political Theory, 36*(3), 377–402.
McDonald, H. P. (2004). *John Dewey and environmental philosophy*. Albany: State University of New York.
Neumann, I. B., & Sending, O. J. (2010). *Governing the global polity: Practice, mentality, rationality*. Ann Arbor: University of Michigan.
Nilsson, A. (2007). *A changing Arctic climate: Science and policy in the Arctic climate impact assessment*. Linköping: University of Linköping, Department of Water and Environmental Studies.
North, D. (1990). *Institutions, institutional change and economic performance*. Cambridge: Cambridge University Press.
O'Neill, J., Holland, A., & Light, A. (2008). *Environmental values*. London: Routledge.
Ockwell, D., Whitmarsh, L., & O'Neill, S. (2009). Reorienting climate change communication for effective mitigation: Forcing people to be green or fostering grass-roots engagement? *Science Communication, 30*(3), 305–327.
Oels, A. (2005). *Theorising power in global climate politics. From bio-power to neoliberal governmentality?* Presentation at the International Studies Association's annual convention in Honolulu.
Okereke, C., & Bulkeley, H. (2007). Conceptualizing climate change governance beyond the international regime: A review of four theoretical approaches. *Tyndall centre working paper 112*. http://www.tyndall.uea.ac.uk/book/export/html/297. Retrieved 17 May 2010.
Olssen, M. (2002). Michel Foucault as "thin" communitarian: Difference, community and democracy. *Cultural Studies, 2*(4), 483–513.

Orlowe, B. (2009). The past, the present and some possible futures of adaptation. In N. W. Adger, I. Lorenzoni, & K. O'Brien (Eds.), *Adapting to climate change: Thresholds, values, governance* (pp. 131–163). Cambridge: Cambridge University Press.

Pellizzoni, L. (2003). Knowledge, uncertainty and the transformation of the public sphere. *European Journal of Social Theory, 6*(3), 327–355.

Pellizzoni, L. (2004). Responsibility and environmental governance. *Environmental Politics, 13*(3), 541–565.

Pettenger, M. E. (2007). *The social construction of climate change: Power, knowledge, norms, discourses.* Aldershot: Ashgate.

Potter, E. (2009). Calculating interests: Climate change and the politics of life. http://journal.media-culture-org.au/index.php/mcjournal/article/view/182. Retrieved 2 Feb 2010.

Reynolds, J. M. (2004). "Pragmatic humanism" in Foucault's later work. *Canadian Journal of Political Science, 37*(4), 951–977.

Roberts, N. (2004). Public deliberation in an age of direct citizen participation. *The American Review of Public Administration, 34*(4), 315–353.

Rose, N. (1996). *Powers of freedom: Reframing political thought.* Cambridge: Cambridge University Press.

Rose, N. (2001). The politics of life itself. *Theory, Culture & Society, 18*(6), 1–30.

Rose, N. (2007). *The politics of life itself: Biomedicine, power and subjectivity in the twenty-first century.* Princeton: Princeton University Press.

Rutherford, P. (2000). *The problem of nature in contemporary social theory.* Ph.D. Thesis. Research School of Social Sciences, Australian National University, Canberra. http://thesis.anu.edu.au/uploads/approved/adt-ANU20011217.114840/public/02whole.pdf. Retrieved 5 Aug 2010.

Schofield, B. (2002). Partners in power: Governing the self-sustaining community. *Sociology, 36*(3), 663–683.

Shadian, J., & Tennberg, M. (Eds.). (2009). *Legacies and change in polar sciences: Historical, legal and political reflections on the international polar year.* Aldershot: Ashgate.

Shields, P. M. (2003). The community of inquiry: Classical pragmatism and public administration. *Administration & Society, 35*(5), 510–538.

Smit, B., & Pilifosova, O. (2003). From adaptation to adaptive capacity and vulnerability reduction. In J. B. Smith, R. J. T. Klein, & S. Huq (Eds.), *Climate change, adaptive capacity and development* (pp. 1–20). London: Imperial College Press.

Smit, B., & Wandel, J. (2006). Adaptation, adaptive capacity and vulnerability. *Global Environmental Change, 16*(3), 282–292.

Staber, U., & Sÿdow, J. (2002). Organizational adaptive capacity: A structuration perspective. *Journal of Management Inquiry, 11*(4), 408–424.

Summerville, J., Adkins, B. A., & Kendall, G. (2008). Community participation, rights and responsibilities: The governmentality of sustainable development policy in Australia. *Environment and Planning C: Government and Policy, 26,* 696–711.

Swanson, D., Barg, S., Tyler, S., Venema, H., Tomar, S., Bhadwal, S., Nair, S., Ro, D., & Drexhage, J. (2010). Seven tools for creating adaptive policies. *Technological Forecasting and Social Change, 77*(6), 924–939.

Sydneysmith, R., Andrachuk, M., Smit, B., & Hovelsrud, G. (2010). Vulnerability and adaptive capacity in Arctic communities. In D. Armitage & R. Plummer (Eds.), *Adaptive capacity and environmental governance* (pp. 133–156). Berlin: Springer.

Tennberg, M. (2000). *Arctic environmental cooperation: A study in governmentality.* Aldershot: Ashgate.

Thompson, K. (2003). Forms of resistance: Foucault on tactical reversal and self-formation. *Continental Philosophy Review, 36,* 113–138.

Tobias, S. (2005). Foucault on freedom and capability. *Theory, Culture and Society, 22*(4), 65–85.

Turnbull, N. (2008). Dewey's philosophy of questioning: Science, practical reason and democracy. *History of the Human Sciences, 21*(1), 49–75.

Turner, B. L., II, Matson, P. A., McCarthy, J. J., Corell, R. W., Christensen, L., Eckley, N., Hovelsrud-Broda, G., Kasperson, J. X., Kasperson, R. E., Luers, A., Martello, M. L., Mathiesen,

S., Naylor, R., Polsky, C., Pulsipher, A., Schiller, A., Selin, H., & Tyler, N. (2003). Illustrating the coupled human-environment system for vulnerability analysis: Three case studies. *Proceedings of the National Academy of Sciences of the United States of America, 100*(14), 8080–8085.

Turnheim, B., & Tezcan, M. Y. (2009). Complex governance to cope with global environmental risk: An assessment of the United Nations framework convention on climate change. *Science and Engineering Ethics*. doi:10.1007/s11948-009-9170-1.

Voβ, J.-P., Smith, A., & Grin, J. (2009). Designing long-term policy: Rethinking transition management. *Policy Science, 42*, 275–302.

Walker, W. E., Rahman, S. A., & Cave, J. (2001). Adaptive policies, policy analysis and policy-making. *European Journal of Operational Research, 128*, 282–289.

Webber, J. (2001). Why can't we be Deweyan citizens? *Educational Theory, 51*(2), 171–190.

Weber, E. P. (2008). Facing and managing climate change: Assumptions, science and governance response. *Political Science, 60*(1), 133–149.

Wilson, H. (2006). Review essay: Environmental democracy and the green state. *Polity, 38*(2), 276–294.

Part II
Russian Adaptation Governance

Chapter 3
Adaptation in Russian Climate Governance

Maria Rakkolainen and Monica Tennberg

Abstract This chapter presents and discusses the practices of Russian adaptation governance. Climate policy in the country has long considered a changing climate to be a natural phenomenon, not a human-caused problem, and it is only recently that this view has started to change on the official level. Hazardous events caused by the warming climate, such as floods and permafrost degradation, as well as the economic losses resulting from such events, are defined as major national and economic concerns. Most of the current governance efforts focus on emergency response; national plans for pro-active adaptation are only just in the making. The chapter discusses a pilot project to develop an adaptation strategy for the Murmansk region. Russian adaptation governance suffers from the same problems as Russian environmental governance in general: a lack of the material, intellectual and organisational resources needed to tackle the issues and implement concrete plans of action.

Keywords Adaptation • Governance • Russia • Murmansk • Federal • Regional • Local

3.1 Russian Climate Governance

Climate change is a complex political and economic issue for Russia. With its extensive Arctic territory, it will experience many of the consequences – both positive and negative – of a warming climate. In particular, easier access to Arctic oil and gas resources and their possible exploitation due to warming have generated a

M. Rakkolainen (✉) • M. Tennberg
Arctic Centre, University of Lapland, P.O. Box 122, 96101 Rovaniemi, Finland
e-mail: maria.rakkolainen@gmail.com; monica.tennberg@ulapland.fi

great deal of interest domestically as well as internationally in the country's Arctic regions. Of note in this regard is that Russia's economy relies heavily on the income from exports of oil and gas, making adaptation to climate change and its impacts a central national strategy; it seems to be more important than the mitigation of harmful emissions. Russia's rationale for its national climate policy is: "We should not choose between emissions reduction and economic development" (Roshydromet 2008; Dobrolyubova and Zhukov 2008, 7).

According to the Russian understanding, the country's role in international climate governance is that of an environmental great power: it played an important role in saving the Kyoto Protocol by ratifying it in 2004, and its forests are a crucial sink for global greenhouse gas emissions (Korppoo 2008, 2009; Tynkkynen 2008). Russia is considered to be one of the countries that could benefit economically from the Kyoto mechanisms, especially emissions trading and joint implementation projects (Kokorin and Safonov 2007, 6; ZumBrunnen 2009, 78). However, the use of these mechanisms in Russia thus far has been limited due to legal and administrative barriers (Korppoo 2008, 2009). The Russian commitment to the post-Kyoto emission cuts (10–15% by 2020) is conditional in that it depends on the commitment of all the major greenhouse gas producing countries.

This chapter presents and discusses the practices of Russian adaptation governance. The first issue to be taken up is the problem of adaptation to the impacts of climate change. Climate policy in the country has long considered a changing climate to be a natural phenomenon, not a human-caused problem, and it is only recently that this view has started to change officially. However, there is no national agreement on how vulnerable Russia is to climate change. Hazardous events caused by the warming climate, such as floods and permafrost degradation, as well as the economic losses resulting from such events, are defined as major national and economic concerns. Russian climate politics in general mostly serves foreign policy and economic needs (Lafel'd and Rote 2007). Most of the current governance efforts focus on emergency response; national plans for pro-active adaptation are only just in the making. Russian adaptation governance suffers from the same problems as Russian environmental governance in general: lack of the material, intellectual and organisational resources needed to tackle the issues and implement concrete plans of action (Oldfield 2005; Tennberg 2007, 2009). While local adaptation plans and policies to deal with the impacts of climate change are rare in practice, one can find some efforts by various societal actors, the media, administration and nongovernmental organisations to start the deliberations such instruments require. A case illustrating such measures in the Murmansk region of the Russian Arctic will be presented and discussed.

3.2 The Russian Problem of Climate Change

Russian scientists have long been quite sceptical of the human role in climate change. Their scepticism is based on two considerations. First, according to the cycle theory, climate change is part of a natural cycle in the Earth's climate; that is, the climate has always fluctuated. The changes have been significant and sometimes even abrupt (Terez 2004, 3, 15; Dobrolyubova and Zhukov 2008, 2, 5). Second, some researchers suggest that climate change is not anthropogenic (Terez 2004, 4, 8; Medvedev 2009; Dobrolyubova and Zhukov 2008). According to this view, changes in solar radiation have not been taken into consideration sufficiently when investigating the causes of climate change (Medvedev 2009; Terez 2004, 12). Moreover, there might be still other factors which could produce climate change and are unknown to science at this time (Volodin 2009, 16).

Despite the scepticism, most Russian studies of climate change focus on the anthropogenic nature of the phenomenon. The researchers believe that the present climate change differs from past climate changes because of the human activities during the last 200 years that have had radical effects on the atmosphere and its composition (Klimenko 2005). The "Climate Doctrine" (Klimaticheskaya doktrina 2009) is the first official document in Russia (now adopted) that considers climate change a result of human activity. Many Russian studies expect that climate change will strongly impact a large number of physical and biological systems. The current estimate is that damage related to climate change already costs Russia up to 1.91 billion dollars per year. Without an economic system for adaptation, the negative impacts of climate change could cost the country as much as 2–5% of its GDP per year (Climate change doctrine signifies policy shift 2009).

Nationally, most worrying are hazardous events related to climate change: According to the Russian strategic prediction (Roshydromet 2005, 8), "growing impacts of hazardous natural events" will contribute to "the increased loss of property and vulnerability of community" (see also Smol'yakova 2009). The most significant economic losses will be caused by hazardous hydro-meteorological events, such as floods. The thawing of permafrost is regarded as the most dangerous phenomenon caused by climate change. Thawing ground is expected to lead to landslides, slower slope flow (solifluction) and significant changes in topography (thermokarst). (Comprehensive Climate Strategy for the Sustainable Development of the Arctic Regions of Russia in the Circumstances of Changing Climate 2009; hereinafter CCS 2009, 29).

Vulnerability is rarely mentioned in political discussions, however. The Russian understanding of vulnerability is quite ambiguous. On the one hand, some official sources (such as Dobrolyubova 2009, 2; Ekonomicheskoe obozrenie LOGOSS-PRESS 2009, 1–2) stress that the country is extremely vulnerable to climate change when compared to other countries. A significant part of the Russian Federation is situated in areas with maximal ongoing and expected climate change (Treyvish 2003; Parshev 2006). Its Arctic regions are seen as being the most vulnerable, because of the vulnerability of ecosystems and socio-economic development in the Arctic (Shestopalov 2008, 1–2).

On the other hand, the Climate Doctrine considers Russia's adaptive capacity to be relatively high compared to that of other countries (Climate Doctrine 2009, 13). This position is supported by invoking the same factors that are seen as making the country very vulnerable to the impacts of climate change. The size of the affected regions and the country's large geographic dimensions contribute to its adaptive capacity. Moreover, only a very small proportion of the population lives in regions highly vulnerable to climate change (ibid., 13–14).

The Climate Doctrine (2009, 13) also identifies certain negative consequences for the Russian Arctic, such as permafrost degradation and the concomitant damage to buildings and communications in northern areas. Positive consequences include less harsh ice conditions and opportunities for cargo transportation in the Arctic seas, facilitating access to the Arctic shelf and its resources. In the Arctic, the primarily negative consequences of climate change will have a considerable impact on ecosystems and socio-economic development. These changes will also affect the life and health of citizens. Indigenous peoples and elderly people are considered the most vulnerable social groups in the context of climate change (ibid., 12).

In Russian politics, the Arctic is seen more in terms of the opportunities afforded by climate change than the threats posed by it. The country's interests in the region are substantial, grounded in history, geography and the resources available there. In the recent Russian Arctic Strategy (2008), President Dmitry Medvedev underlines that the Arctic must be the country's "top strategic resource base" by 2020 (Emerson 2007; Russia's Arctic Strategy 2008). From this perspective, in political terms the opportunities for maritime transportation, tourism and access to oil and gas resources outweigh the threats to indigenous peoples and other vulnerable groups in the Russian Arctic.

In general, the Russian understanding of climate change as a problem of adaptation seems quite contradictory. Rowe (2009; see also Forbes and Stammler 2009) describes the Russian problematisation of climate change as "domesticated" in that it incorporates elements from the international climate science debates with some key domestic beliefs, albeit somewhat "ambivalently". The domesticated problem of climate change includes competing explanations for the warming of the climate, ones in which climate change has only recently been officially accepted as a human-induced problem. Moreover, there is no agreement on how vulnerable Russia is to the impacts of climate change. The same reasons are cited for considering it either very vulnerable or very adaptive. The only agreement on the problem of adaptation seems be on the threat of hazardous events and on their economic importance. Climate change will increase the hazards related to flooding and permafrost degradation, leaving infrastructure and people vulnerable (Roshydromet 2005, 9–12).

3.3 Federal Adaptation Governance in Russia

In connection to the national Climate Doctrine, an additional plan of adaptation has now been adopted (CCS 2009; Kompleksnyy plan realizatsii Klimaticheskoy doktriny 2011). The Climate doctrine (2009, 22) introduced a new concept, "preventive adaptation", which refers to "measures to adapt to climate change, including preventive adaptation, are planned and implemented in the framework of public climate policy taking into account sectoral, regional and local context and the long-term character of these measures, their influence on different aspects of social and economic and public life". National adaptation governance is in the planning stage. For national climate politics, the focus is on developing adaptive strategies and practical measures for adaptation to climate change rather than on reducing emissions. This emphasis on adaptation instead of the mitigation of greenhouse gas emissions has attracted a great deal of criticism, especially from nongovernmental environmental organisations (e.g. Sheyermeer 2009; Ozharovskiy 2009; Kireeva 2009).

The role of the Russian state in adaptation governance so far focuses on risk assessment and management. Adaptation governance is mostly reactive, dealing with hazards after the fact. Russia is very much focused on dealing with "acute serious problems" on both the national and regional levels (See Yanitzky 2000; Crotty and Crane 2004). Where emergency response is concerned, the central actor is the Ministry of Russian Federation for Civil Defence, Emergencies and Elimination of Consequences of Natural Disasters (EMERCOM). The Russian Rescue Corps was established in 1990 in order to respond to emergencies and to provide safety for those in different parts of the country. In co-operation with the Russian Academy of Sciences, EMERCOM has analysed probable threats, such as natural disasters, accidents caused by technology and environmental changes. EMERCOM has 33 state-funded units, with 3,700 members of staff at eight search and rescue centres. In addition, there are more than 350 locally funded emergency and rescue units in different parts of the country whose staff totals some 10,500. (Porfiriev 2001; EMERCOM 2010)

In regard to national adaptation plans and strategies, Aleksey Kokorin (2009) argues that "the big chink in Russian climate policy is the lack of efforts which could promote adaptation." The development of adaptation governance in Russia is difficult since there is no coherent climate policy at either the national or regional level (e.g. Adaptatsiya 2009; Ozharovskiy 2009). With no unified state agency, body or structure that concentrates on developing the lines of climate policy in Russia, the governance of climate change is fragmented and unclear. Among the agencies responsible for climate policy at the moment are the Ministry of Foreign Affairs (international processes), the Ministry of Natural Resources (environmental affairs), the Department of Energy (energy), the Ministry of Finance (economic affairs), the Economic Development Department (Kyoto projects) and the Ministry of Emergency Situations (hazards).

Also playing a key role is the Federal Service for Hydrometeorology and Environmental Monitoring (Roshydromet), which is responsible for monitoring, assessment and scientific studies. The main documents of Roshydromet dealing with climate change and its impacts are the *Strategic Prediction of Climate Change Expected in Russia for the Period 2010–2015 and its Impact on Sectors of the Russian National Economy* (2005) and *The Assessment Report on Climate Change* (2008). The plans for adaptation in Russia are currently in the research and planning phase. Some large-scale research projects geared to the development of risk management methods (e.g. modern installations for Roshydromet) have now started with government funding.

There is a problem with financial resources for adaptation in Russia. First, it is challenging to combine evaluations of adaptive strategies with cost-benefit analyses of financial viability in order to develop profitability measures for adaptive strategies. Some Russian experts point out that the costs of adaptive measures are most often higher than those of repairing the most serious consequences of the problem or even eliminating the problem (Dobrolyubova 2008).

Second, there will be little funding for adaptation and adaptive programmes on the national level. Implementation of adaptive measures within the framework of the separate environmental national programme is almost impossible because of the general lack of financial resources for environmental programmes. Financial support has to be arranged from the local budget or sources of external funding. Russian federal investments in adaptation take place only in projects which are profitable or might have strategic importance. In addition, the Russian government is interested in external sources of financing (Review of international experience in adaptation of cities to the climate change and further perspectives for developing adaptive strategies for Moscow City 2009; CCS 2009).

One concern in the future for Russian climate research will be human resources for adaptation governance. The limited understanding of climate change among civil servants and stakeholders is also an obstacle to the development of adaptation governance. Climate change appears to be too much of an "ecological" problem for the Russian government to deal with. Civil servants have difficulties in recognising the economic aspects of climate change and its impacts. The administration of environmental protection in general is not attractive for the recruitment of young, talented people; many of them end up working for international organisations and projects instead. (Review of international experience in adaptation of cities to the climate change and further perspectives for developing adaptive strategies for Moscow City 2009, 53). It should also be noted that there is a wide range of other serious and current economic and social problems that compete for attention with climate change. Moreover, discussions of the positive impacts of climate change have delayed the development of adaptation plans and strategies (Uyazvimost' i adaptatsiya 2009). What is more, political inaction in developing adaptive strategies is based on an assumption that the prevention of problems related to climate change is not particularly relevant because of the nature of the problem, which is uncontrolled, unpredictable and could change over time (Dobrolyubova 2008; CCS 2009).

Russian adaptation governance currently focuses on emergency preparedness and response. Plans for adaptation are only just in the making. The plans and strategies for climate change adaptation suffer from a lack of organisational, material and human resources. To some extent, this might be explained by the Russian tradition of environmental measures, which is more reactive than preventive. In the case of adaptation measures, there are a number of governmental actors, one of the most important being Roshydromet, while for emergencies the role of EMERCOM is central. In many respects, adaptive strategies in Russia have not yet reached the practical level. To date, the priorities of the adaptive climate policy of the Russian Federation have been the development of a regulatory framework and financial mechanisms for adaptive measures, international co-operation in adaptation, adaptation research and human resources development (Climate Doctrine 2009, 14–15).

3.4 Regional Adaptation Governance in Russia

According to the main Russian documents on climate change and adaptation (e.g. Climate Doctrine 2009; CCS 2009), adaptation should be promoted on the national, regional and local levels by taking into account what will be unevenly distributed impacts of climate change. According to the recently published Climate Doctrine, the particular conditions in Russia should be taken into account in plans for adaptation. The first of these conditions is the combination of the low population density and large size of regions affected, which leads to increased transport costs for citizens and challenges the infrastructure that is essential for the economy. A second consideration is the cold climate, which imposes some additional requirements where the heating of buildings and houses is concerned. A third concern is how to produce and transport large volumes of fuel and energy resources. (Climate Doctrine 2009, 14)

Bringing a regional perspective to adaptation studies is an important shift in climate policy in Russia. A regional focus is a new way to understand adaptation and climate change in the country, especially from a political point of view. Some efforts at regional adaptation governance started in 2008 in the Murmansk region with the organising of an international conference on adaptation to climate change. The Murmansk case study (CCS 2009) is a pilot study and a first project for the region in the area of adaptation to climate change in Russia. The conference served as a forum for exchanging practical experiences on the elaboration of regional adaptation strategies and on their financing, and the strategies have been introduced into regional development programmes (Final Statement of the Participants of the International Conference 2008).

The Murmansk Region has a relatively high GRP (gross regional product, the regional analogue of the gross national product). The region has important industrial, transport and energy sectors, as well as an extensive fish-processing industry, and the importance of ecotourism has grown recently. The sector most vulnerable to climate change in the Murmansk region is agriculture, which is highly vulnerable to extreme natural events (e.g. wind, heat, and rain). However, the impacts of climate change on agriculture could well be positive: The growing season could become longer and precipitation in the region could increase (CCS 2009, 35). The conference also contributed to discussions about actions and priorities in the Murmansk region related to climate change, including the identification of the region's most vulnerable groups. The population groups most vulnerable to climate change impacts in the region are older people (about 13% of the population) and indigenous peoples (ibid., 17).

One challenge for regional adaptation governance is that there is a very limited amount of regionally relevant information and few studies on the impacts of and adaptation to climate change. In particular, there is a lack of research evaluating the economic losses caused by climate change. The documents for the Murmansk case draw attention to the unpredictability of climate change effects (CCS 2009, 26, 33). More detailed studies are needed on how climate change will impact the Arctic regions.

The most important task of science nowadays appears to be the development of regional climate change models that make it possible to formulate detailed regional climatic predictions and the analysis of that information from an economic point of view. For example, in the case of Murmansk there are Russian environmental organisations, mainly supported by external foreign funding, that produce local knowledge and improve environmental awareness (e.g. Russian Regional Environmental Centre, Bellona, WWF).

The Murmansk case study points to two interactive strategies that should be elaborated to ensure sustainable development in Russia's Arctic regions: a strategy for (1) the traditional economy of nature areas, and for (2) active areas of industrialisation. Both strategies must take into account the interests of indigenous peoples (See CCS 2009, 7). The document states that the Murmansk region could be developed by using its fuel and energy resources, exploiting its scientific potential and commercialising tourism services (ibid., 14). At the regional level, adaptation is seen as an integral part of regional social and economic development strategies. Developing regional action plans and adaptive strategies in collaboration with federal authorities, enterprises, the media, institutes and communities is also important. The role of municipalities is strengthened on the regional level as part of the implementation of national and regional target programmes and environmental measures (Russian progress report 2007, 40–41).

In the Murmansk documents, adaptive strategies are considered to be long-term options for the regions. The Murmansk strategy (CCS 2009, 44–45) points out two different approaches to the realisation of adaptive measures: a regional approach (districts, areas, regions, transnational regions, e.g. co-operation within Barents region and the EU) and a sector-drawing approach based on economic sectors, such as agriculture, infrastructure, health, or on societal groups, such as indigenous and elderly people. The conference also furthered the use of evaluation as a means to develop adaptive strategies. This created a basis for additional regional studies on adaptation to climate change. The Murmansk case study (CCS 2009) examined concrete regional adaptive measures and regional priorities, including the economic and industrial sectors. Even though adaptive measures have a declarative nature, the measures to be implemented for the Murmansk region are more multifaceted. According to CCS (2009, 44–45), adaptation should include regional and national strategies as well as practical measures on the individual or community level. Finally, the conference discussed the establishment of a permanent consulting body that would focus on adaptive strategies in the Murmansk region (Materialy konferentsii i kruglogo stola po voprosam adaptatsii k izmeneniyu klimata Murmanskoy oblasti 2008).

The conference was organised by the United Nations Development Programme (UNDP) and the Russian Regional Environmental Centre with support of the Roshydromet, the Administration of Murmansk oblast, and the Institute of Industrial Ecology of the North at the Kola Science Centre of the Russian Academy of Sciences. The forum of some 60 participants at the conference included representatives of federal authorities of the Russian Federation, the Administration of Murmansk oblast and its municipal areas, regional businesses, as well as experts from the Russian Academy of Sciences, the Faculty of Geography of the Moscow State University, the scientific and research institutes of Roshydromet, various Russian public bodies, and international organisations such as the UNDP, European Commission, the Regional Environmental Centre for Central and Eastern Europe, and the World Bank. In Murmansk, the region is being developed in accordance with the Russian Federation's outline of socio-economic development for the region until 2020, a document which contains concrete adaptive measures (see Table 3.1).

In Murmansk, there is also international co-operation in the area of emergency response. EMERCOM (2003) suggested to the Arctic Council in 2003 that an Arctic rescue mechanism be developed to deal with emergencies in the region. In the context of the Arctic Council, a new treaty on search and rescue in the Arctic was signed in May 2011. The legally binding treaty will improve co-ordination among Arctic countries in the event of a plane crash, cruise ship sinking, major oil spill or other disaster. The need for such a treaty came from the opening of Arctic waterways to more marine traffic, including shipping vessels and cruise ships, due to melting sea ice. In addition, international search and rescue exercises have been organised in the Barents region, the most recent being that held in 2009 in Murmansk. Cross-border co-operation on emergencies is important since in the case of an emergency the closest rescue resource might be located on the other side of a national border. For example, an oil spill in Norwegian waters could easily turn into a Russian problem and vice versa. International co-operation is therefore important and will be even more important in the future if tourism and cargo shipping in the Barents region increase as anticipated. Most of the international co-operation in the Barents region takes place in the border area between Norway and Russia. The recent search and rescue exercises were carried out in the waters of Kola Bay to prepare for various situations, such as a collision between a shuttle tanker and oil tanker that leads to an oil spill. The rescue services of Norway, Russia and Sweden worked together to deal with such an emergency (Barents observer 2009).

3.5 Local Adaptation Governance in Russia

Local administration in Russia has been transformed considerably since the 1990s (Toschenko and Tsikov 2006; Heusala 2005; Didyk 2010), but its role in the Russian adaptation governance system is clearly still undetermined. There is very little local action taking place where adaptation governance is concerned. The Murmansk pilot study (CCS 2009) concluded that there is a need for local knowledge and skills related to the impacts of climate change and adaptation at the local level. According to the pilot study (CCS 2009), local authorities could play an important role in promoting adaptation. Local governments should have a clear picture of the impacts and intensity of climate change on the regional scale. Based on such knowledge, adaptation should be promoted particularly at the regional and local levels in accordance with the natural and socio-economic characteristics of each region. Here a

Table 3.1 Adaptive planning in the Murmansk region

	Murmansk region
Concrete adaptive measures	Only in the planning phase; (e.g. measures, related to assessments and reporting, use of water)
Detailed studies of the region	Murmansk (2008) international conference on adaptation to climate change and its role in ensuring the sustainable development of regions
	Murmansk I international economic forum (2009)
	Comprehensive climate strategy for the sustainable development of the Arctic regions of Russia in the circumstances of changing climate (Moscow 2009)
Existing regional documents that include adaptive measures	The Russian Federation's concept of socio-economic development of the region until 2020
	According to the concept, the following adaptive measures should be implemented in region:
	1. Development of oil and gas field exploitation
	2. In 2011, geological and geophysical studies that substantiate and consolidate the external borders of Russia's continental shelf should reach completion
	3. Building of new railway lines and roads in order to ensure cargo transportation in the Arctic regions
	4. Modernisation of Murmansk harbour and development of interregional connections (e.g. improving main routes)
Barriers to adaptation and developing of adaptive strategies for the region	1. No monitoring system available
	2. Financial problems
	3. Lack of political will
	4. Problems with regulations
	5. Lack of a system for assessing economic and environmental risks
	6. Inaction by and lack of confidence in executive bodies and authorities (these were mentioned for the Murmansk region in particular)
What should be done in order to eliminate the barriers?	1. Investments should be made in the development of a monitoring system
	2. A comprehensive federal programme and a number of federal regulations should be created (adaptation as a part of national and regional strategies)
	3. The number of training courses and information sessions should be increased for both decision-makers and citizens;
	4. A comprehensive federal adaptive programme should be developed
	5. Co-operation at different levels of government should be strengthened, taking into consideration climate experts, civil society and business representatives
	6. International co-operation and exchange of experience should be promoted

Source: CCS 2009; Materialy konferentsii i kruglogo stola po voprosam adaptatsii k izmeneniyu klimata Murmanskoy oblasti (2008); Final statement of the participants of the international conference "Adaptation to climate change and its role in ensuring sustainable development of regions" (2008); Kontseptsiya dolgosrochnogo sotsial'no-ekonomicheskogo razvitiya RF na period do 2020 g (2008); Russia's Arctic Strategy (2008); Review of international experience in adaptation of cities to the climate change and further perspectives for developing adaptive strategies for Moscow City (2009)

key role will be played by the regional and local administrative bodies as well as the structures responsible for the development and design of the regions.

This view is supported by the adaptation research. According to a study by Keskitalo and Kulyasova (2009, 242), the role of governance in community adaptation to climate change is central. Drawing on case studies of fishing in Northwest Russia, they show that adaptive capacity beyond the immediate economic adaptations available to local actors is, to a considerable extent, politically determined within larger governance networks. The adaptive range of local communities is limited by the institutional framework: more far-ranging adaptations on the local level would require changes at the level of state regulation on resource access and other support. According to international climate governance, adaptive measures should be decided at the appropriate level of government, strengthening co-operation between different levels of government, climate experts, civil society and business representatives (CCS 2009, 44–45).

In general, the Russian population is aware of climate change but not very concerned about it. An opinion poll from 2006 (Kommersant 2006) showed that 63% of Russians believe that the ecological situation in the country is becoming worse. The majority of the population (86%) was more worried about water pollution than about climate change. According to a more recent study, conducted in 2009 (Krechetnikov 2009), environmental problems are not generally at the top of the list of citizens' main concerns and serious problems. Only 2–3% of citizens regarded environmental problems as their chief concern. Eleven per cent cited climate change as being among other important environmental problems.

However, Russians know about climate change. In opinion poll carried out in summer 2008, (see Ivanova 2009) the majority of respondents were aware of the phenomenon (86%). Two-thirds believed in it (67%) and 15% thought that it is not real. Eighteen per cent of respondents were not able to assess the situation. Climate change seems to be a very abstract concept for ordinary citizens in Russia. The risks related to climate change are not apparent to them and they are more concerned about concrete and common problems related to environmental protection, such as water pollution.

It seems that climate change is far from people's daily life. In political statements one can very often find arguments such as "There is enough time to act by 2050 before something bad happens" or "Now it's better to concentrate on serious problems that we already have and that need to be solved" (e.g. Gavrilov 2007.) People are not able to relate their personal interests to their community and to the generations that will follow them. Moreover, a majority of Russians think that environmental issues are a special interest for a "narrow" group.

There is a special mentality about the future in Russia, one based on "surviving day to day". A wide range of serious economic, political, governmental and environmental problems has created different understandings of what "serious problem" means in the country. For Russians, "serious" is something concrete and very close – happening now, not in 2050. This focus on daily survival can be attributed in large measure to uncertainties and the unstable political and economic system (Kommersant 2006).

3.6 Conclusions

International climate governance is based on an economic logic, on cost-benefit analysis and the use of market mechanisms to steer human activities. However, this particular economic rationality seems to work poorly in the Russian domestic political and economic context. The Russian governmental responsibility for adaptation to climate change is fragmented, reactive, and limited. At the moment, Russian adaptation governance appears to be very centralised, a top-down exercise with intellectual, material and organisational resources that are too limited to allow it to be properly developed and implemented. Russian adaptation governance is reactive and corrective, focusing on natural hazards and responses to them; a proactive approach would include planning to support and advance sustainable development in different parts of the country.

Russian adaptation governance suffers from are weak signals and a lack of knowledge between different governmental levels and bodies. Regions and local communities have not emerged as a site of adaptation governance. The Murmansk region appears to be a case study, a laboratory, for developing Russian adaptation governance in collaboration with international partners. The development of Russian adaptation governance is important for the future of the country's Arctic region. The Soviet era produced heavy industrial structures and the settlement patterns to support them, and some of the projects extended into remote Arctic regions. The significance of the Arctic has been highlighted with a spirit like that of a "conquering state" and with an emphasis on the economic importance of the region (Ria "Novosti". Goryachaya liniya 2009; see also about Russian policy in the Arctic: Razuvaev 2006; Problemy malochislennykh narodov Rossiyskogo Severa 2007; Riabova 2010). However, it is uncertain how those remote regions, peoples and livelihoods can be supported and maintained in the future. The future challenge of adaptation in Russia is considerable: How can the country develop adaptive plans and strategies that support development in different regions and direct resources from profitable, mainly natural-resource-based activities, to "unprofitable, but vital sectors and regions" such as its Arctic and Siberian settlements? (Lynch 2005, 222).

References

Adaptatsiya (2009). (Adaptation). Informatsionnyy portal "Global'noe izmenenie klimata". http://www.climatechange.ru/node/371. Retrieved 2 Nov 2009.

Barents observer (2009, Dec 9). Barents rescue 2009. http://www.barentsobserver.com/barents-rescue-2009-.4630842-116320.html. Retrieved 16 Aug 2010.

Climate change doctrine signifies policy shift (2009). Oxford Analytica Daily Brief Service. http://proquest.umi.com/pqdweb?did=1739396601&Fmt=3&VInst=PROD&VType=PQD&RQT=309&VName=PQD&. Retrieved 16 Aug 2010.

Climate doctrine project (2009). Klimaticheskaya doktrina Rossiyskoy Federatsii. http://www.mnr.gov.ru/files/part/9500_project_climate_doktrine.doc. Retrieved 1 Aug 2011.

Comprehensive climate strategy for the sustainable development of the Arctic regions of Russia in the circumstances of changing climate (CCS) (2009). Kompleksnye klimaticheskie strategii dlya ustoychivogo razvitiya regionov rossiyskoy Arktiki v usloviyakh izmeneniya klimata, Model'nyy primer Murmanskoy oblasti. Moskva: UNDP & RRETS. http://www.rusrec.ru/files/Murmansk_report_sm.pdf. Retrieved 16 Nov 2009.

Crotty, J., & Crane, A. (2004). Transitions in environmental risk in a transitional economy: Management capability and community trust in Russia. *Journal of Risk Research, 7*(4), 413–429.

Didyk, V. (2010, Dec 14). Formation of local self-government and administrative reform in Russia: Aims and reality – case of the Murmansk region. Presentation at the BIPE workshop, Rovaniemi, Finland.

Dobrolyubova, Yu. (2008). *Vysokiy gradus rossiyskogo severa* (More intense clime in the Russian North). Informatsionnoe agenstvo "Rosbalt-Sever" 17/05. http://www.rosbalt.ru/2008/05/17/484476.html. Retrieved 2 Oct 2009.

Dobrolyubova, Yu. (2009). *Tayushchaya krasota. Izmenenie klimata i yego posledstviya* (Melting beauty. Climate change and its impact). Rossiyskiy regional'nyy ekologicheskiy tsentr Fond im. Genrikha Byollya. http://www.climatechange.ru/files/RREC_Boell_Melting_Beauty.pdf. Retrieved 1 Aug 2011.

Dobrolyubova, Yu.C., & Zhukov, B.B. (2008). *10 samykh populyarnykh zabluzhdeniy o global'nom poteplenii i Kiotskom protokole* (10 most popular misunderstandings about global warming and the Kyoto protocol). Russian Regional Environmental Centre: Moskva. http://www.rusrec.ru/files/RREC_Brochure_Top10.pdf. Retrieved 20 Dec 2009.

Ekonomicheskoe obozrenie LOGOS-PRESS (2009). *Vsemirnyy bank predosteregaet* (The world bank warns).12.6, 22 (806). http://aafnet.integrum.ru/artefact3/ia/ia5.aspx?lv=6&si=AmHkak2R&qu=231&st=0&bi=7299&xi=&nd=1&tnd=0&srt=0&f=0. Retrieved 5 Dec 2009.

EMERCOM (2003). Arctic rescue. Initiative proposed by the EMERCOM of Russia. http://eppr.arctic-council.org/pdf/ArcticRescue.pdf. Retrieved 16 Aug 2010.

EMERCOM (2010). EMERCOM of Russia. http://www.mchs.gov.ru/en/. Retrieved 16 Aug 2010.

Emerson, C. (2007). Russia has reasons to turn up the heat in the Arctic. *Financial Times* Apr 16, 2010, p. 9.

Final statement of the participants of the international conference "Adaptation to climate change and its role in ensuring sustainable development of regions" (2008). http://www.rusrec.ru/files/Final%20statement_eng.doc. Retrieved 12 Dec 2009.

Forbes, B., & Stammler, F. (2009). Arctic climate change discourse: The contrasting politics of research agendas in the West and Russia. *Polar Research, 28*, 28–42.

Gavrilov, V. (2007). *Voyny iz-za tepla. Eksperty OON predrekayut global'nye konflikty na pochve izmeneniya klimata* (War on heat. UN experts predict global conflicts caused by climate change). RBK daily. Rossiyskaya akademiya nauk. http://www.ras.ru/digest/showdnews.aspx?_language=ru&id=0765f7f1-654d-4644-8b41-b01fcca117c6. Retrieved 22 Nov 2009.

Heusala, A-L. (2005). *The transitions of local administration culture in Russia.* Kikimora Publications A:12. Helsinki: Gummerus.

Ivanova, M. (2009). *Byt' bolee razumnymi* (To be smarter). Nezavisimaya gazeta. Rossiyskaya akademiya nauk 29/9. http://www.ras.ru/digest/showdnews.aspx?_language=ru&id=35195740-74ed-428e-bfe5-9d9d9d1188c3. Retrieved 2 Aug 2011.

Kireeva, A. (2009). *Prezidium pravitel'stva RF rassmotrel Klimaticheskuyu doktrinu RF* (RF Government Presidium working on the climate doctrine of Russian Federation). Klimat. Bellona 07/05-2009. http://www.bellona.ru/articles_ru/articles_2009/1241686219.67. Retrieved 8 Dec 2009.

Keskitalo, E. C., & Kulyasova, A. (2009). Local adaptation to climate change in fishing villages and forest settlements in Northwest Russia. In S. Nystén-Haarala (Ed.), *The changing governance of renewable natural resources in Northwest Russia* (pp. 227–243). Aldershot: Ashgate.

Klimenko, B. (2005). *Global'nye izmeneniya klimata: Chto zhdet Rossiyu* (Global warming: What is in store for Russia?). Polit.Ru. http://www.polit.ru/analytics/2005/01/12/klim.html. Retrieved 22 Oct 2009.

Kokorin, A. (2009). *V Rosii prinyata Klimaticheskaya doktrina* (Climate doctrine adopted in Russia). WWF 24/4. http://www.wwf.ru/resources/news/article/4996. Retrieved 1 Dec 2009.

Kokorin, A. O., & Safonov, G. V. (2007). *Chto budet posle kiotskogo protokola?* (What comes after the Kyoto Protocol?) Mezhdunarodnoe soglashenie ob ogranichenii vybrosov parnikovykh gazov posle 2012 g. Vsemirnyy fond dikoy prirody, WWF Rossii. http://climategroup.org.ua/upl/WWFpost-2012ru.pdf. Retrieved 27 Nov 2009.

Kommersant (2006). *Besformennoe bespokoystvo: problemy okruzhayushchey sredy volnuyut zhiteley razvivayushchikhsya stran, no ne vosprinimayutsya kak problema politicheskaya* (Formless concern: environmental problems disturb people in developing countries, but are not perceived as a political problem). 28.4.2006. www.kommesant.ru/Doc-rss/670504. Retrieved 1 Dec 2009.

Kompleksnyy plan realizatsii Klimaticheskoy doktriny (2011). Comprehensive plan of implementing the Russian Federation's climate doctrine for the period until 2020). government.ru/media/2011/4/29/40950/file/730R_pril.doc. Retrieved 3 Aug 2011.

Kontseptsiya dolgosrochnogo sotsial'no-ekonomicheskogo razvitiya RF na period do 2020 g (2008). The Russian Federation's concept of socio-economic development of the region until 2020. http://www.rusnanonet.ru/download/nano/conception_2020.pdf. Retrieved 24 Aug 2008.

Korppoo, A. (2008, Nov 24). Russia and the post-2012 climate regime: Foreign rather than environmental policy. UPI briefing paper 23. http://www.upi-fiia.fi/en/publication/61/. Retrieved 10 June 2011.

Korppoo, A. (2009, June 5). The Russian debate on climate doctrine: Emerging issues on the road to Copenhagen. UPI briefing paper 33. http://www.fiia.fi/en/publication/79/the_russian_debate_on_climate_doctrine/. Retrieved 10 June 2011.

Krechetnikov, A. (2009). *Rossiyane o global'nom poteplenii: nam by Vashi zaboty* (Russians on global warming: Could we have your worries?). BBC. Russkaya sluzhba.13/10. http://www.bbc.co.uk/russian/russia/2009/10/091012_russia_climate_change_attitude.shtml. Retrieved 1 Nov 2009.

Lafel'd, S., & Rote, Yu. (2007). *Rossiya: Energeticheskaya politika i uglerodnyy rynok* (Russia: Energy policy and carbon markets). 3C climate change consulting GmbH. http://www.wwf.ru/data/pub/russiaandcarbonmarket_rus.pdf. Retrieved 1 Aug 2011.

Lynch, A. C. (2005). *How Russia is not ruled. Reflections on Russian political development.* Cambridge: Cambridge University Press.

Materialy konferentsii i kruglogo stola po voprosam adaptatsii k izmeneniyu klimata Murmanskoy oblasti (2008). Conference papers for the roundtable on adaptation and climate change in the Murmansk region. RRETS. http://www.rusrec.ru/ru/news/1513. Retrieved 11 Oct 2009.

Medvedev, Yu. (2009). *Klimaticheskiy triller: Chelovek ne vinovat v global'nom poteplenii* (Climate thriller: man is not guilty of global warming). Rossiyskaya gazeta 04/08. http://www.rg.ru/printable/2009/08/04/poteplenie.html; http://www.rg.ru/2009/08/04/poteplenie.html. Retrieved 12 Nov 2009.

Oldfield, J. D. (2005). *Russian nature: Exploring the environmental consequences of societal change.* Aldershot: Ashgate.

Ozharovskiy, A. (2009). *Razgovor s prezidentom o climate* (A conversation with the president on climate). Klimat Kiotskiy protocol. Bellona 20/04-2009. http://www.bellona.ru/comments/Medvedev-climate. Retrieved 5 Dec 2009.

Parshev, A. R. (2006). *Pochemu Rossiya ne Amerika* (Why Russia is not America.). Krymskiy most-9D, Forum. http://lib.ru/POLITOLOG/PARSHEW/parshew.txt. Retrieved 16 Jan 2010.

Porfiriev, B. (2001). Institutional and legislative issues in emergency management policy in Russia. *Journal of Hazardous Materials, 88*, 145–167.

Problemy malochislennykh narodov Rossiyskogo Severa (2007). *Narodonaselenie* (Settling). 31/03. http://aafnet.integrum.ru.login.ezproxy.ulapland.fi/artefact3/ia/ia5.aspx?lv=6&si=AhYJph2R&qu=221&st=0&bi=5600&xi=&nd=1&tnd=0&srt=0&f=0. Retrieved 22 Nov 2009.

Razuvaev, P. (2006). *Bor'ba. Krasnyy Sever* (Struggle, Beautiful North). 25/09. http://aafnet.integrum.ru.login.ezproxy.ulapland.fi/artefact3/ia/ia5.aspx?lv=6&si=AhYJph2R&qu=241&st=0&bi=6831&xi=&nd=3&tnd=0&srt=0&f=0. Retrieved 22 Oct 2009.

Riabova, L. (2010, December). State policy in the Russian North, its social outcomes and needs for change. Presentation at the BIPE workshop, Rovaniemi, Finland.

Review of international experience in adaptation of cities to the climate change and further perspectives for developing adaptive strategies for Moscow City (2009). Obzor mezhdunarodnogo opyta v oblasti bol'shikh gorodov k klimaticheskim izmeneniyam i perspektivy razrabotki strategii adaptatsii dlya goroda Moskvy. Proekt "Klimaticheskie strategii dlya rossiyskikh megapolisov". Fond strategicheskikh programm Ministerstva inostrannykh del Velikobritanii. http://russian-city-climate.ru/adaptation_review.pdf. Retrieved 1 Aug 2011.

Ria "Novosti". Goryachaya liniya (2009). *Rossiya dolzhna kak mozhno skoree nachat' adaptatsiyu ekonomiki k izmeneniyu klimata* (Russia should start to adapt the economy to climate change as soon as possible). VB. 24/06. http://aafnet.integrum.ru.login.ezproxy.ulapland.fi/artefact3/ia/ia5.aspx?lv=6&si=fuEiFb2R&qu=231&st=0&bi=1198&xi=&nd=1&tnd=0&srt=0&f=0. Retrieved 1 Dec 2009.

Roshydromet (2005). *Strategic prediction for the period of up to 2010–2015 of climate change expected in Russia and its impact on sectors of the Russian national economy* (Strategicheskiy prognoz izmeneniy klimata Rossiyskoy Federatsii na period do 2010–2015 gg i ikh vliyaniya na otrasli ekonomiki Rossii). http://www.meteorf.ru/rgm3d.aspx?RgmFolderID=8fa3a439-2cb4-4d09-b567-36fd11f3f414&RgmDocID=71a57c11-f042-47d7-89b1-7b1c57933f50. Retrieved 26 Jan 2010.

Roshydromet (2008). *Assessment report on climate change and its consequences in Russian Federation* (Otsenochnyy doklad ob izmeneniyakh klimata i ikh posledstviyakh na territorii Rossiyskoy Federatsii). General summary. Moscow. http://www.meteorf.ru/default_doc.aspx?RgmFolderID=a4e36ec1-c49d-461c-8b4f-167d20cb27d8&RgmDocID=7bd005e1-a689-485c-9b95-aa0a71dcc2be. Retrieved 26 Jan 2010.

Rowe, E. W. (2009). Who is to blame? Agency, causality, responsibility and the role of experts in Russian framings of global climate change. *Europe-Asia Studies, 61*(4), 593–619.

Russia's Arctic strategy (2008). Vyezdnoe soveshchanie po strategicheskomu planirovaniyu "O zashchite natsional'nykh interesov Rossiyskoy Federatsii v Arktike" 13.9.2008. http://www.scrf.gov.ru/news/349.html. Retrieved 7 Nov 2009.

Russian progress report (2007). Doklad ob ochevidnom progresse v vypolnenii obyazatel'stv Rossiyskoy Federatsii po Kiotskomu protokolu. Moskva. http://unfcc.int/resource/doc/dpr/rus1.pdf. Retrieved 15 Nov 2009.

Shestopalov, M. (2008). *Vektor ustremleniy – Arktika: problemy ustoychivogo razvitiya arkticheskoy zony Rossii* (The direction of aspirations – The Arctic: Challenges for sustainable development of the Russian Arctic). Geopolitika i geostrategiya 10/11. http://aafnet.integrum.ru.login.ezproxy.ulapland.fi/artefact3/ia/ia5.aspx?lv=6&si=TOfEdM2R&qu=221&st=0&bi=5480&xi=&nd=1&tnd=0&srt=0&f=0. Retrieved 22 Nov 2009.

Sheyermeer, K. (2009). *Perestroyka i vechnaya merzlota: Moskva nachinaet proyavlyat' interes k izmeneniyam klimata* (Perestroika and permafrost: Moscow begins to show interest in climate change). Inoforum 28/5. http://inoforum.ru/inostrannaya_pressa/perestrojka_i_vechnaya_merzlota_moskva_nachinaet_proyavlyat_interes_k_izmeneniyam_klimata. Retrieved 27 Oct 2009.

Smol'yakova, T. (2009). *Priroda nakazhet rublem: Yesli segodnya ne brat' v raschet izmeneniya klimata, zavtra eto privedet k snizheniyu VVP* (Nature will punish the ruble: If we do not take into account climate change today, tomorrow it will cause the GDP to decrease). Rossiyskaya gazeta Ekonomika " Ekologiya" 4938 (112) 23/06. http://www.rg.ru/printable/2009/06/23/trutnev.html. Retrieved 11 Dec 2009.

Tennberg, M. (2007). International environmental cooperation in Northwest Russia: An assessment of performance. *Polar Record, 43*(226), 231–238.

Tennberg, M. (2009). Regional governance, path-dependency and capacity-building in international environmental cooperation. In S. Nystén-Haarala (Ed.), *Changing governance of renewable natural resources in Northwest Russia* (pp. 245–257). Aldershot: Ashgate.

Terez, E. I. (2004). *Ustoychivoe razvitie i problemy izmeneniya global'nogo klimata zemli* (Sustainable development and problems of global warming). Mezhdistsiplinarnaya ploshchadka *poteplenie.ru*. Tavricheskiy natsional'nyy universitet im. *Vernadskogo, 17*(56), 181–205. http://www.poteplenie.ru/doc/terez.htm. Retrieved 15 Oct 2009.

Toschenko, Z., & Tsikov, T. (2006). The formation of democracy and self-governance in Russia. In S. Szücs & L. Strömberg (Eds.), *Local elites, political capital and democratic development governing leaders in seven European countries* (pp. 234–255). Wiesbaden: VS Verlag für Sozialwissenschaften.

Treyvish, A. (2003). *Nasha strana samaya kholodnaya v mire* (Our country is the coldest in the world). Znanie-sila. 5/03, Sankt Peterburgu trista let: I eto vse Rossiya. http://www.znanie-sila.ru/online/issue2print_2159.html. Retrieved 8 Dec 2009.

Tynkkynen, N. (2008). Constructing the environmental regime between Europe and Russia. Conditions for social learning. http://acta.uta.fi/pdf/978-951-44-7268-8.pdf. Retrieved 16 Aug 2010.

Uyazvimost' i adaptatsiya (2009). Vulnerability and adaptation. Informatsionnyy portal "Global'noe izmenenie klimata". http://www.climatechange.ru/node/3. Retrieved 5 Dec 2009.

Volodin, E. (2009). *Dostovernost' prognoza budushchikh izmeneniy* (Reliability of the prediction of future changes). http://83.149.207.89/GCM_DATA_PLOTTING/documents/forecast.pdf. Retrieved 1 Aug 2011.

Yanitzky, O. N. (2000). *Russian greens in a risk society: A structural analysis.* Kikimora Publications B:11. Helsinki.

ZumBrunnen, C. (2009). Climate change in the Russian North: Threats real and potential. In E. W. Rowe (Ed.), *Russia and the North* (pp. 53–85). Ottawa: University of Ottawa Press.

Chapter 4
The Big Water of a Small River: Flood Experiences and a Community Agenda for Change

Anna Stammler-Gossmann

Abstract This chapter draws on observations from anthropological fieldwork in a flood-prone area in northeast Siberia to comment on how rural residents assess and "process" destructive consequences of floods in their interaction with water and multiple water governance scales. The massive Soviet "river-turning" projects figure prominently in the development programmes in the Tatta District, an area that has suffered from drought for several decades. Due to the state programme for transporting water from the Lena River, the largest in northeast Siberia, to Tatta, the small Tatta River became a dense network of water management projects. The research demonstrates that the governmental and local agenda for dealing with increased flooding – the causes of such a departure from the norm, the view of "things to be governed" and "how they should be governed" – may differ considerably. This paper focuses on how rural communities search for a balance in their adaptive practices amid several tensions: conflicting attributions of disaster to dams and irrigation constructions as opposed to natural changes; the threat of increased flooding alongside a practical need for development projects; and the clash between the governmental 'emergency' approach and communities' long-term adaptive practices.

Keywords Affordances • Culturally "affiliated" community • Russian North Republic of Sakha Yakutia • Tatta River • Flood • Natural disaster • Development projects • Human-nature interaction • Emergency strategy • Perception and action Local adaptation

---

A. Stammler-Gossmann (✉)
Arctic Centre, University of Lapland, 122, 96101 Rovaniemi, Finland
e-mail: anna.stammler-gossmann@ulapland.fi

4.1 Introduction

Until recent times, residents of Tatta Ulus (the Tatta District in the Republic of Sakha Yakutia in northeastern Siberia, Russia; hereinafter Tatta) have seen themselves as people inhabiting a stable environment where the principal threats are natural stresses. The river network on the Tatta plain is the lowest not only in Sakha (Yakutia) but in the entire boreal forest. For agro-pastoral communities in Tatta located in the basin of several rivers, dealing with "wet" and "dry" years is nothing new. The Sakha expression for the cyclic "wet" years (*ugut djyllar*) has the positive connotation of "wet years beneficial for the growth of plants and grass". The most common outcome of "wet" years is flooding of grazing areas, hay grounds and agricultural land. These types of changes are known to the locals and are addressed in daily practice. They have maintained a balanced relationship with their water environment.

On the regional planning agenda, the Tatta District has belonged to the areas suffering from drought for several decades. The Soviet utopian idea of "taming nature" by "diverting" rivers from the European North and Siberia to the southern regions of the country and by other projects designed to redistribute water resources (Micklin 1987; Vorobyev 2005) was well represented in the regional development programmes related to the Tatta River. As a result of these projects, the small river became a dense network of water management efforts. In 1992–1993, the Sakha government introduced a programme designed to transport water from the Lena River, the largest in Northeast Siberia, to a local lake and from there to the Tatta River (Decree No. 256, 15.07.1992; Salva 1999). The main goal of this project was to support local agriculture through melioration of fields and hay grounds. Around 60 earthen water storage dams and two sluice gates were built for irrigation purposes in Tatta and the neighbouring Churapcha District.

However, in the last decade the Tatta region has appeared on the regional agenda as one of the areas most exposed to the greatest flood risk. Spring floods have become one of the most frequent hazards in the Republic of Sakha Yakutia, which has one fifth of the rivers in Russia. According to the newly established regional "Plan for the protection of settlements and economic objects of the Republic of Sakha Yakutia from flood" (Decree No. 253, 27.05.2010), the heaviest floods in Russia have taken place in this region. The large as well as the small rivers of this northern region, like the Tatta River, cause considerable problems for rural areas. The agro-pastoralist settlements located in the Tatta River valley – Ytyk Kyöl (the administrative centre of the district) and nearby small villages – experienced exceptional heavy flooding in 2007 (see Figs. 4.1, 4.2).

The increased frequency of severe floods in the different regions of the republic in the last decade and in Tatta in 2007 and 2008 has placed questions on the local communities' agenda about negotiating their position in a new, changing environment. Events of such incredible magnitude have challenged local norms and practices. People have lost essential places – not merely homes and infrastructure, but gathering sites, formal public hubs, and environmental features that convey a

Fig. 4.1 The Tatta River in Ytyk Kyöl (Photo: Anna Stammler-Gossmann)

Fig. 4.2 The big water of a small river: flooding in Ytyk Kyöl in 2007 (Photo: Alexander Postnikov)

sense of community and identity. Evacuation, displacement projects, and the threat of the next potential flood have brought into the foreground issues of attachment to place and building their milieu anew. Issues of law, legality and insurance have arisen, as have questions of resource sharing and inequality.

One factor that challenges the residents of Tatta is that flooding events and the uncertainties related to them occur side by side with the "production of wealth". Adaptation-related proposals are placed among those serving people's practical needs to have water for melioration, drinking water and transport infrastructure and to address concerns over increasing environmental degradation caused by intensive river engineering, large deforestation, and a failure to recultivate land after road construction.

This chapter investigates how the inhabitants of Tatta, for whom uncertainty has become a measurable feature of the past, assess and "process" the destructive consequences of floods in the context of interaction with water and multiple water governance scales. The theoretical discussions on phenomena deemed to be risks concentrate mainly on projected future danger or, as Beck puts it, "second-hand non-experience" (Beck 1992, 72); yet, very little is known about the power of the past to determine the present. There is no firm grounding in empirical research on how the disaster-development relationship is conceptualised and understood in contemporary society in the context of climate change on the global agenda. The relationship between natural and anthropogenic changes also raises some difficult questions. Since it is uncertain how the world climate would change in the absence of landscape engineering, should we continue to shape nature according to our concepts of "progress" and "prosperity"? How is it possible to plan for the unpredictable and the indeterminate? How can we find a balance when adjusting to natural fluctuation that occurs without human input and to human-induced changes without setting back development?

In this chapter, I draw on observations from anthropological fieldwork to comment on the nature of a community's perception of extreme events – floods – and its participation in the process of governing those events. Analysis on the local scale affords insights into the community's course of action in the context of a "one size fits all" national policy proposal for water governance and flood management. I argue that a community's ability to respond to change in the most positive, constructive ways depends on an understanding of what "a thing to be governed" is and how that should be governed. In doing so, the chapter draws attention to the interaction of Tatta residents in both the physical and social environment – the activities of people as they engage in the world. This inquiry may help to understand which factors foster, mandate, and shape the human response to hazardous events and facilitate people's adaptive agency.

4.2 Reflections on Human Beings and the Environment

A strong conceptual separation – in Hettinger's words a "human/nature apartheid" – exists between humans and nature (Ingold 1992; Hettinger 2005; Heyd 2007). Previous studies have pointed to the difficulties of applying this idea to capture how local people address changes in their environment (Forbes and Stammler 2009), to envision a positive role for humans in nature in preservationist thoughts (Hettingen 2005; Heyd 2007) or to make it easier to control the state of the environment (Berger 1998, 2008).

The findings of my research on the Russian state's approaches to flood management demonstrate that governmental policy continues to be dominated by the notion of valuing humans and nature separately. For the most part, floods have been governed at the institutional level as a "natural disasters". In the Russian governmental context, the term "disaster" is generally employed to characterise not a process as Oliver-Smith defines it (Oliver-Smith 1999, 2002), but an event (see Law N 68-FZ, 21.12.1994; Roshydromet 2005). Approaching a disaster as a rather surprising anomaly determines the means of responding to the event. My analysis of national and regional legal documents and fieldwork observations shows that "emergency" is the principal concept for coming to terms with "natural hazards" in Russia. The national mitigation industry is concerned with flood control, often defined in terms of flood elimination. It does not necessarily comply with local practices. Records of Tatta residents, as well as examples from my other research site in Russia (Stammler-Gossmann 2010a), illustrate that local people regard the abnormal imperatives of their physical environment in long-term perspective rather than as exceptions that justify the rules.

Proceeding from conceptual insights on human-nature relations as an interaction within the same world (see Ingold 1992, 2000), I extend the notion of the mutualism of human beings and nature to the "social frame" of interaction and develop an approach for community initiatives that unfold in interaction with the immediate and "societal" environment. Ideas of affordance, or use-value, from ecological psychology offer some guidance and ways of thinking about human involvement and interaction with multi-entities. In this vein, our immediate perception of the environment is not a passive sensing, but occurs in terms of what the environment affords for the action in which we are currently engaged (Gibson 1977, 1979). Specifically, the view of affordances that I adopt in this chapter is one of practical possibilities for action or opportunities for a community experiencing a flood in the world around it.

This study is based on the data collected during intensive fieldwork conducted in 2008 and 2009 in two rural settlements in the Tatta District: Ytyk Kyöl (6,952 inhabitants, municipal archive 2007) and Uolba (some 500 inhabitants). I used a number of complementary methods, including participant observation, oral history, in-depth interviews and media research. The fieldwork was carried out among residents of private cattle-breeding households, village residents, and representatives of the municipalities and the local mass media. Participant observation, which involved living with host families of horse and cow breeders, made it possible to get important information about ongoing changes and how these changes are dealt with in everyday life in hunting, fishing, and travelling. In the regional capital of Yakutsk, I interviewed researchers from the Permafrost Institute, Sakha State University, the Arctic Institute, the Institute of Engineering and Construction and the Academy of Agriculture, as well as persons working in the regional administration and ecologists.

4.3 Case Study Area

Tatta Ulus is one of the 34 administrative municipal districts of the Republic of Sakha Yakutia (see Fig. 4.3). Cow and horse breeding and agriculture are the main economic activities in this rural area. Different kinds of intimate convertible knowledge and skills and informal social networks provide the basis for the flexible response that rural residents employ to adjust to the complexities of environmental and social changes. Following the economic and social transformations of the post-Soviet period, which included the collapse of collective farms, very low "added value" milk and meat production and institutional restructuring, many people in the villages of Tatta have drastically reduced their milk consumption; they need cash and sell almost their entire milk production to the processing plant in the summer time. In winter they sell meat and frozen milk in the district centre. As the price paid by the state for milk products is not satisfactory, people actively use their personal networks to sell the products.

Shifts in the structures in which one participates are common practices among breeders in maintaining control over their private economies. Depending on which cow-breeding structure receives more subsidies, people may switch between the Agricultural Production Co-operative (the quasi-state enterprise), the Farmers' Co-operative and the recently introduced Individual Farm Enterprise. Associated membership in co-operatives may also be only formal and in reality people may have other jobs in the budgetary system. Some people may consider breaking their dependence on the mainstays of the local economy: "If I could no longer keep my animals, I would probably move to the town and start doing construction work" (personal conversation 2008).

Species diversification on farms, intensive gardening and use of locally available wild food resources have been the most common practices to overcome the challenges of the transition period from a centralised planned economy to a market economy. Adjusting local decisions may include a strategy of maintaining one's herd capital through animal redistribution or increased mobility and accessing more distant hay grounds (see Fig. 4.4). Diversifying the economy helps a community's members minimise the damages caused by downturns in cow and horse breeding or environmental changes. Changes accepted as a part of the "natural order" may be perceived as manageable with regard to their consequences: "When our hay grounds are flooded, we can use relatives' grounds or rent the grounds from the municipality". Those villagers who have limited opportunities for economic diversification invest more in the elaboration of their social network. This involvement of people in their environment in the practical context of picking up information, moving around and negotiating changes is a mode of engagement with the world around them.

Although Tatta communities share the same experiences with rural communities elsewhere in adjusting strategies to the economic, social and environmental changes, the district has some significant distinctive features. First, it is considered one of the best-known agricultural areas in Sakha and the "most Sakha" province (Sakha is a self-designation of the northernmost Turkic-speaking ethnic group in

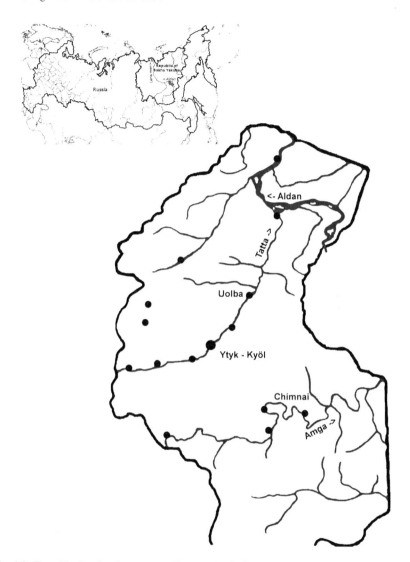

Fig. 4.3 Tatta District: flood prone area (Source: Arctic Centre)

northeastern Russia; "Yakut" is the ethnonym in Russian). Meat and butter from Tatta, as well as the region's potatoes and cabbage, are renowned for their high "organic" quality and are highly valued in the markets and among individual consumers in the regional capital Yakutsk. This image is supported by the clean environment and high grass quality of the district's *alaas*, fertile meadows for grazing sites and hay grounds. My research partners in Uolba explained the quality of their milk and meat products with reference to the particular distribution of this highly valued land in Uolba: Each household in the village was allocated 2 ha of grazing and hay land per person and per head of cattle, while in other settlements this

Fig. 4.4 Regular flooding of pastures: part of a seasonal cycle in the agro-pastoralist community of Uolba (Photo: Anna-Stammler-Gossmann)

number might be only 0.5 ha. In addition to its allocated land, my host family rented another 20 ha of land in 2008 for its seven cows, four calves and unspecified number of horses. The availability of and access to grazing land and hay grounds are a main issue for the cow breeders of Tatta during both "dry" and "wet" years: "In these periods cows go too far away trying to find good pastures and do not come back home. We can search for them for days and they produce less milk under these conditions" (personal conversation 2008).

The preparation of hay is crucial in cattle breeding in northern districts. Around 2.5 tonnes of hay are needed for one cow in Tatta to survive the harsh winter. The additional artificial feeding extensively used in the Soviet period is not valued among herders. It would be a heavy burden on the family budget and might undermine the reputation of animals grazing on Tatta *alaas* by changing the taste of the meat and milk products. Beyond their economic significance, attachment to *alaas* is an important component of personal identities and how people feel about their community. Moreover, the symbol of the *alaas* is part of the Sakha ethnic identity and may provide a supportive potential for Sakha "ruralism" in regional politics (Stammler-Gossmann 2010b; Adamov 2010).

Another specific feature of Tatta is that the region belongs to the core areas of the cultural revival of the Sakha people. It is from there that the Sakha intelligentsia derives its roots, energy and emotional ties. It is a homeland of the founders of the Sakha national literature, famous philosophers, musicians and artists. The residents of Tatta carefully support the museum culture and even small settlements like Uolba have an impressive museum collection on the history, culture and nature of the area.

As people belonging to a culturally "affiliated" community (Rose 1996, 340), Tatta residents have the educational and moral means to "pass" in their role as active citizens. A particular kind of investment in themselves and in their families can be observed in the rearing of children, in schooling, training, and other impressive efforts in educational and cultural activities. Authorities who came to Ytyk Kyöl during the flood events were asked for support not only to rebuild damaged houses and infrastructure, but also to establish a chess centre, a Centre for National Food and a branch of the regional university. The lack of access to the capital when the road infrastructure was destroyed caused great alarm in the district, as seen in media reports and in residents recalling the events during my fieldwork. The disruption caused difficulties in obtaining the documents for the unified state exam in time, which Tatta school children needed to get a place at university.

Tatta's glorious historical accounts reveal considerable effort on the part of the community to participate in different local and regional educational and cultural projects. In almost all the households I visited, I got to admire all possible awards for the best conference presentation, performance, national dress, school essay, participation in environmental projects, handicraft project, and the like. When the people found out I was a researcher, I was asked to suggest some new interesting research topic that they could study for the next local or regional project. Issues related to environmental changes are well integrated into the school curricula and were presented outstandingly in the school essays that I had an opportunity to read during my fieldwork in 2008–2009.

At the same time, Tatta belongs to the most silent areas in the region when it comes to political thought. During the Soviet time, Tatta Ulus carried the label of the "cradle of Sakha nationalism". A decision of the Central Committee of the Communist Party in 1928 on Sakha nationalism brought widespread repression among the Tatta intelligentsia. Echoes of this were heard in the discussions from the 1980s and resulted in a widespread negative impact on Tatta's residents. After a publication of the KGB (Soviet Secret Service) in 1985 about a deeply rooted anti-Soviet movement in Tatta during World War Two (Alekseev 1996: 101–104), the district became an object of special political control. Educational barriers, such as limited access to university, reduced subsidies for schools and public housing, or reduction of the main Sakha festival *ysyakh* to a sporting event linger in the memory of the community. Its morality and lifestyle were considered a threat or a rebuke to the political order in the Soviet period. Many people who now have key a decision-making role are still familiar with the consequences of this time. The position of Tatta Ulus between its "marginalised" and "affiliated" status finds its reflection in communities' engagement in cross-scale interactions.

4.4 Perception of the Flood: Technological Failure or Natural Hazard?

The flood of 2007 – in contrast to previous and subsequent flood events – was severe not only in magnitude but also in duration. Two of Tatta's settlements were flooded completely and in the district's flooded administrative centre, Ytyk Kyöl, the water level exceeded the critical level by 72 cm and stayed above it for 1.5 months (Municipal archive 2007). The flood destroyed all bridges in the village and all municipal roads and paralysed transportation throughout the district. The federal road that connects Tatta to the Lena River – the only route in the region to the republic capital, Yakutsk, and the only connection to the outside world – was destroyed as well (see Fig. 4.5).

Domestically, events such as floods are rarely interpreted as manifestations of climate change. The Russian situation can be described as reflecting the "climate paradigm". The political and academic interest in climate change that has started to grow in the country gains its motivation primarily from the international community and the greater integration of Russia into global discourses. For example, the Russian "equivalent" of the ACIA, "Climate change impact on public health in the Russian Arctic", (Revich 2008) is a result of international co-operation and a product of experts working within the UN system in Russia. Until recent times, the signals from federal to regional authorities on the issue have been very weak. Public pressure to act is low and climate scepticism remains high. The fieldwork in 2008–2009 indicated that the regional perception, as seen in official reports and the mass media, hardly ascribes the flood to climate change. The issue of climate change is not on the community's agenda either. Should one ask about climate change, the locals may redirect the question to a scientific field of concern.

Residents' perceptions of floods encompass a multitude of factors, but cluster around irrigation dams, water reservoirs and hydro-technical constructions built for the as yet unfinished water supply project. The state project on water governing culminated in 2004 when a 149-km canal being built from the Lena River reached the Tatta River in the neighbouring district. National and regional information agencies reported cheerfully about the filling of 53 lakes in and around the district with more than 40 million cubic metres of water (Regnum.ru 2004). The Decree "On irrigation and water supply" proposes a new state programme that should continue construction of local lines to the village of Uolba and seven new sluice gates (including four in the Tatta District) (Decree GS No 98-III, 25.04.2003).

The most extreme floods in the area – those in 1980 and 2007, when 89% and 60% of Ytyk Kyöl, respectively, disappeared under water – have been identified by downstream users as the result of technological failure and unintentional human error. Dam projects have exacerbated certain tensions between different categories of local users. Villagers put the blame for the flood on their Churapcha neighbours, who "did not open sluice gates in time". Overfilling of water reservoirs placed along the river and the rupture of earthen dams in Ytyk Kyöl that brought a flash flood are among the key points in a 2007 municipal report addressed to the regional

Fig. 4.5 Federal road in the Tatta District: the only connection with the outside world and one of the causes of the 2007 flood (Photo: Alexander Postnikov)

government and several appeals by the local administration to the regional president, Parliament and different governmental units (Municipal archive 2007).

Among the causes of flood events residents also identify the intensive deforestation and logging activities of the local lumber mill in which wood waste was stored on the completely dried out river bed during the almost 10-year drought. The construction of a federal road by the end of the 1990s that blocked the water outflow also figures in explanations of the flood. While pastoralists may tolerate partial flooding of their hay grounds as "nature's course", human-induced changes may be perceived as more disturbing.

In the view of communities in Ytyk Kyöl and Uolba, forest infestation by bark beetles that killed several hectares of the local forest is a result of using pipes from China with beetles nesting inside. The impacts of dust pollution of grazing land, soil erosion and boggy landscape caused by gravel roads, forest fires and deforestation have been framed as "disastrous" along the same lines as floods. These changes create stress, disrupt normal social processes, and force people to make temporary or permanent adjustments in how and where they do things in life. Local people's priorities and their individual evaluation of the causes and effect of natural disasters are important resources in understanding how Tatta communities work and shape their attitudes toward changes.

The regional way of comprehending severe flood events in the area falls within the national hazard framework, in which the main approach is to see disastrous events as applied to the society in economic terms. In this case the magnitude of the disaster is categorised in terms of damage to goods, property and infrastructure (Law No. 68-FZ, 21.12.1994). As the field work material illustrates, this "economic" classification can have significant implications for regions like Sakha with its vast territory; it is the size of India but has a population of only around one million, a very low population density, small settlements and inadequately developed infrastructure. During flood events, these characteristics have influenced the flow of financial assistance and shaped the limits of regional flood management. Many houses in Tatta, especially summer cottages, do not have complete documentation. The federal centre has not acknowledged buildings without documentation as objects of compensation.

On the regional agenda, the economic definition of the flood cannot be fully attributed a primary role. The natural disaster has been mainly explained in terms of "accidental" geophysical features of a place. As the regional minister of the Ministry of Emergency Situations reported in 2006, the main cause of the floods in Sakha Yakutia has been spring ice jams (Sukhoborov 2006). Indeed, the rivers of the region that flow from the south to the north can bring an intensive snowmelt from the warmer upper reaches, resulting in an extremely high inflow of water and formation of powerful ice jams downstream. Rain and snow were confirmed as reasons in the case of the heavy flooding of agro-pastoral land in 2005 in Tatta. In 2007, when the government officially designated the situation in the district as "extreme", the main cause of the flood was recognised as rapid melting of snow.

Among the shifting broad array of "objective" natural factors that produce or trigger disaster and that "could not be prevented" (Sukhoborov 2006), technological phenomena appear as secondary factors. In the case of Tatta, engineering complications – when not wholly ignored by the government – are separated from environmental issues or are too segmented and compartmentalised. This makes it more challenging to put together a coherent account of governmental development strategies, the impact of floods and environmental sensitivities.

4.5 State of Emergency

Approaching disasters as departures from the norm and emphasising their apparent brevity determine the means of responding to the events. Calhoun accounts for the construction of "emergencies" in terms of a social imaginary that gives characteristic form to both perception and action (Calhoun 2004). The focus of Russia's emergency policy is the use of remedial, curative measures using best available techniques. The national structure for co-ordinating and executing disaster risk reduction was established in 1992. Since 1994, the Russian Federation's Ministry of Emergency Situations has been responsible for the implementation of disaster management policies and belongs to one of the most effective national institutional systems of emergency response (EMERCOM 2004; APN 2005).

Several laws related to emergency situations have been adopted by the Russian government (WCDR 2005; ADRC 2006). Intervention focuses mainly upon

4 The Big Water of a Small River…

Fig. 4.6 Emergency help: EMERCOM team during the flood in the Tatta District (Photo: Alexander Postnikov)

emergency rescue, rehabilitation of the population and reconstruction of livelihoods once the crisis occurs (see Fig. 4.6). Saving lives or alleviating suffering in emergencies is understood to be immediately good, and emergency operations become highly prized for their media coverage. Nowadays, patterns of national policies are gradually shifting from immediate response to floods towards more integrated approaches to the issues underlying the emergency. Several revisions of a 1994 federal law titled "Risk reduction and mitigation of consequences of emergency situations caused by natural and technogenic factors" have highlighted the importance of action to reduce impacts of extreme events not only as they occur, but before they occur as well (Law N 68 FZ, 21.12.1994).

The "technogenic" component of the new definition of "emergency situation", though it downscales the "man-made" aspect, has brought more holistic elements into disaster management since the last decade. Floods are now more likely to be seen as hazards that have to be controlled through monitored, predictive and preventive measures. However, immediate response and technical effectiveness remain the key components of the emergency strategy in which the government puts its main effort. This means that much less in the way of financial resources have been allocated to comprehensive non-crisis flood risk reduction, but that more have been earmarked for intervention.

Since the frequency of floods in the region has been on the increase, the Sakha government has been urged to intensify flood prevention activities. The republic has adopted hundreds of decrees and resolutions related to floods. More than 290 official documents were introduced in 2001 alone after the flood in Lensk, when images of a town with a population of 28,000 people disappearing under water were broadcast on the world news. The regional programme "Risk reduction and mitigation of consequences of emergency situations" (Z N 217-II, 06.08.2000; Z N 512-III, 16.06.2005) put special emphasis on establishing forecasting, monitoring, and the mobilisation of information retrieval services institutionally. A new institution, the Water Directorate of the Lena Basin, a regional branch of the Ministry of Natural Resources, started its activities monitoring the use of water resources (Decree No. 726, 25.09.2000). In 2007, the Centre for Monitoring Emergency Situations in the Republic of Sakha Yakutia was established (Decree N 143, 12.04.2007).

In the dominant regional geophysical interpretation of the causes of flooding, the "emergency" approach means an accumulation of preventive practices, above all those for controlling ice jams. Different methods are applied to regulate jam formation and river flows: explosives, darkening ice with coal, sawing ice, and widening the river's flow through drilling and digging. In 2007 alone, 39 explosions were carried out on the ice fields of the small Tatta River and intensive widening of the river bed took place as well.

Another widespread practice is to appeal to the population for humanitarian assistance. It is the most "attractive face" of emergency situations. However, it is not necessarily applied on a voluntary basis. Mandatory transfer of money – usually 1 day's salary (indirectly effected in the case of private enterprises) – to help the regional government fund operations is a common way of acquiring extra resources.

4.6 Whose Responsibility?

Approaching a flood as an emergency has made it difficult for the local communities to trace responsibilities. The new revised version of the federal law "On risk reduction" delegates many functions in emergency management to regional and local governance (Law N 206 FZ 18.12.2006). It divides emergency situations into federal, regional and municipal categories. This categorisation determines the sources of financial assistance and responsibility for emergency management. Thus, the question "Who will pay?" is placed between levels in the administrative hierarchy: Should funds come from local, regional or central budgets?

Flood insurance schemes are very rudimentary and the only insurance company operating at the time of the flood refused to issue policies for the impacted areas. On the local level, this situation brought the introduction of a quota system by the regional government, with some ten policies distributed in Ytyk Kyöl after the flood in 2007. The local people complain about extremely stringent criteria in damage assessment: "To get compensation, the water level has to reach almost the roof of the houses".

Uncertainties in identifying the causes of flood contribute to uncertainties in the question of responsibility for concrete adaptation measurements. In summarising the results of work done after the flood in Tatta in 2007, the vice-chair of the Sakha government stated that floods cannot be managed by planned preventive programmes, which require huge investments under conditions where the government is unwilling to take responsibility for the wrong application of funds: "Who will be responsible when for example we have invested money in the river-bed-cleaning programme to prevent floods and the cause proves to be permafrost processes?" (Alekseev 2007).

As conditions worsen and resources become more limited, regional and local administrations are forced to transfer their decision-making to and become more reliant on state government. The common strategy in the region to get support from Moscow is to apply for categorisation of a disaster event as a federal-level emergency. However, according to the governmental "Classification of emergency situations caused by natural and technogenic factors", in order to prove achieve this categorisation, "the number of people impacted by the flood has to be more than 500, and the economic damage should be more than 500 million roubles" (Decree RF No. 304, 21.05.2007). This practice is also adopted on the local municipal level, but it is not an easy task, especially for the rural areas with small settlements and inadequately developed infrastructure. Several applications from different areas in Sakha (Yakutia) affected by floods have been not accepted. With around 900 heavily impacted households in its largest settlement, Ytyk Kyöl, the 2007 flood in Tatta received federal status and the best emergency practices were brought to bear during the disaster of this year.

The local administration tried to secure diverse funds for the emergency situation of 2007. Thirty housing certificates were allocated from the State Emergency Fund; another 30 families were resettled into new houses using regional funds. The acceptance of the flood as a federal emergency enabled authorities to include the Tatta administrative centre in a federal project on assistance for housing and a power line reconstruction project. At the same time, the Tatta residents who were most severely affected in the district, for example those in the small neighbouring village of Chimnai (around 600 inhabitants), discovered significant gaps in the emergency assistance outside of the federal category. The resettlement programme for Chimnai, which had experienced several floods in recent years, was a main issue on the regional government's agenda. After 2007, the relocation plan was abandoned due to budget deficits. Instead of relocating the village, a new milk processing plant was built in the community in 2010 with regional funds.

The emergency imaginary encourages an image of sudden, unpredictable and short-term phenomena, whereas the reality commonly involves production of "the good in some longer term fashion" (Calhoun 2004, 389). A key challenge for the local population lies in building a bridge between current disaster risk management efforts and efforts geared towards longer-term adaptation. Being crowded in between these two forms of decision-making – in Beck's definition, "organised irresponsibility" (Beck 1999, 149) – where there seems to be no institution specifically responsible, challenges the perceived value of action in a community.

All water resources in Russia are state property, and river embankments, for example, are a federal responsibility. Plans by the regional branch of the Ministry of

Emergency Situations to counter disasters proceed separately from the processes operating development plans, where different governmental units are responsible for exploitation of water resources. Dam management is a function of the Ministry of Agriculture and Ministry for Communal Services and Energy.

The construction of a pipeline in Tatta connecting the Tatta and Lena Rivers was financed mainly by regional funds. According to federal law, the safety of dams is the duty of the one who owns and operates them (Federal Law No 117-FZ 21.07.1997). Safety control is a state responsibility. Monitoring of water resources is a task of the Committee for Natural Resources Protection. Luckily, the difficult debate on whose responsibility it was to rebuild the dyke in Ytyk Kyöl found a solution when the federal government acknowledged that the dyke was owned by the municipality. This allowed the local administration to apply for federal support (Decree RF No.522, 29.06.2010).

The question of responsibility is closely intertwined with societal views and collective-cultural constraints. The experience of living many years under the conditions of the totalitarian system in the Soviet Union, as well as the turbulence of the post-socialist period, created a duality of social trust-non/trust in Russian citizens with respect to state actions. On the one hand, people still believe that the state is obligated to help them in technologically induced risks and rely on its action. Local communities emphasise their low capacity to control and respond to "technogenic" disasters without the emergency assistance framework and state reimbursement (Municipal report 2007).

On the other hand, communities do not trust that the state is able to provide them with stability in their day-to-day lives and bring into balance the challenges of development projects and environmental risks. The silo approach – where the existing water management is incapable of reciprocal operation with other sectors within the system and communication is vertical – reduces the perceived value of the state's responsibility. In Tatta, with its memory of marginalisation, trust-non/trust is also a "fear" duality (Marková and Gillespie 2008). Reliance on authorities' power is still high, but in the absence of strong public associations – blocked under the totalitarian regime – people in Tatta assume that their public initiatives may bear the stamp of their past legacy, leading them to avoid any conflict with or direct action towards state authorities. Locals recall the marginalisation of Tatta in the Soviet era and may say about themselves that they are "too shy and politically not active" or point out that "we are not so well represented in the government and the few high-ranking officials from Tatta are also rather silent".

4.7 The Use-Value of the Emergency Frame

"The good" in a short-term emergency arrangement may not resonate with the local frame of long-term oriented adaptation strategies. The use-value of a purely managerial approach may be perceived as low. The particular focus of an "emergency" framework on the disaster phase does not provide direct forms of assistance for preparing before and recovering after the disaster phase. Governmental compensation for damages is never enough and is always viewed as being far below the amount needed to rebuild a livelihood. Administratively introduced flood mitigation requirements may even undermine awareness of the scale of hazardous changes: "We had so many meetings before and after the flood that we somehow lost a feeling of the real danger" (personal communication 2009).

In the year after the flood of 2007, the regional mass media were wondering why it was precisely in the heavily impacted communities that the government's preventive measures in establishing hydrological stations – work which should be carried out by the local municipality – did not function properly (Levochkin 2008). One of the reasons was that while the government emphasised scientific methods for predicting floods, the technological measures were not accompanied by non-technical actions and were not considered as appropriate warning signals by the local administration and residents.

According to a survey carried out in Ytyk Kyöl by the school project on the flood, 46% of local residents did not believe that a flood would occur in spite of the warning (Lopatina 2009). The low credibility of the outside warning sources in combination with "unknown" risk factors influenced the perception of risk when the water level was already rising: "The water came with the speed of a galloping horse; we did not manage even to prepare our flight properly" (personal conversation 2008).

During the flood, citizens also ignored to the extent possible orders to evacuate their property. Forecasting and early warning systems are often the weakest element in the chain of measures designed to reduce the risk of flood disasters. One may describe the spring flood events in Sakha Yakutia by saying that floods always occur where they are not expected. Past technological measures that were designed to prevent flooding, such as strengthening the sand dyke in Ytyk Kyöl, helped in 2005 but did not work in 2007. As often happens in the Russian context, the emergency approach initiates the main activities when a disaster is already underway. The large amounts of money spent on the dyke project are seen by locals as a waste of resources.

Interruption in the implementation of governmental projects due to funding shortages is a reality that makes the "institutionalising" aspect less reliable in the local view. The shortcomings of reliance on the state support became especially obvious in Tatta when the company in charge of elevating the dyke in Ytyk Kyöl during the flood stopped working because it had not received payment from the government on time. The programme to improve the drinking water supply started in 2003 has been paralysed because of funding shortages. People consider the construction of sluice gates or maintaining of the old hydrological facilities to be a state responsibility, but do not believe that the government will be able to do it before the next disaster occurs or do it properly (see Fig. 4.7).

However, the generalised distrust and the common view that residents themselves cannot influence state policy are not necessarily only a detrimental factor for local action. The attitudes may block local initiatives but may also provide them with flexibility. In their proactive adaptation, people attempt to "sum up" various kinds of ambiguities into a local, practical locus. The interplay between diverse elements of

Fig. 4.7 Failed project: water pipeline in Uolba (Photo: Anna Stammler-Gossmann)

ambiguity and uncertainty may produce innovative recombination, akin to that in Stark's account on organisational ecology describing the "cross-fertilising" qualities of friction between multiple standpoints (Stark 2009, 18). In other words, Tatta communities may benefit from the interaction between multiple, sometimes incompatible, principles for their strategies without competing with the state or nature.

Pretended trust can function as a strategy of protecting or enhancing a community's own interests. Rather than oppose state policy, villagers prefer to act legally and peacefully through mediating local officials. The local administration remains an essential actor in villagers' social network. The heroic and competent behaviour of municipal leaders during the flood increased the trust in the local leadership of Ytyk Kyöl tremendously. The quality of local leadership is cited as an important factor in the ability of communities to get things done. However, the interrelationship between village residents and local authorities is very contextual, and residents of another village reported very little trust in their leader. When the legal methods available become insufficient, people attempt to mobilise their personal contacts in the higher governmental institutions.

As disasters occur at the interface of society, technology and environment (Oliver-Smith 2002), the interaction between different connected categories is crucial for what I would call here "proactive" or "creative" agency. People living on the Tatta River try to make assets of diverse evaluative frameworks in their mode of adaptation. Keeping ambiguity in play is not simply a search for beneficial opportunities; it offers connections between traditional ties, local knowledge, relations of trust/distrust, different social logics, and residential and family life.

Local practices of adjustment become more exposed to new challenges, but are mobilised within a heterogeneous spectrum of resources for agency – knowledge, people, money, and information. In creative action, agro-pastoralists of the Tatta region combine common practices of close monitoring of their environment and economic diversification with an exchange of information about new state regulations in agriculture and strategic considerations of what structure is the most beneficial for their enterprises financially. They keep formal and informal networks intact, but search for possibilities to expand their value chain beyond close bonding.

The social assets of kinship ties and personal networks open secure avenues for decision taking under conditions perceived to be increasingly unfavourable. This is not a new strategy, but the investment in long-term social relations becomes more intensive. Adaptation to uncertainty is associated with precautions that may, for example, be placed within generalised reciprocity among relatives.

During the flood in 2007, people experienced new possibilities to extend social networks to include more loosely connected and distant associates. In doing so, they utilised an "institutionalised" form of the state emergency strategy, a humanitarian approach. Local medical workers applied to medical institutions outside of Sakha for help and got substantial support. The amount of financial support for Tatta families from the Far East, 70 thousand roubles, was definitely larger than the state assistance, which amounted to 20 thousand roubles per household. The same approach was undertaken by the library's staff and electricians. One individual initiative became public, and was "enforced" by the Ministry for Health Care through existing network linkages: "I worked at that time in the hospital and asked somebody in the Ministry whom I know to help us" (personal conversation 2009). In this way Tatta residents turned their passive role as recipients or beneficiaries of humanitarian aid into one of more creative social agency.

Shortages in state capacities and late implementation of intended plans in elevating the dyke in Ytyk Kyöl unexpectedly had a "positive outcome". On the one hand, non-involvement in the decision-making process and an absence of public associations may undermine human agency at the personal level. On the other hand, the issues relating to the dyke created an opportunity for collective agency. When the water level rose and the road construction company was not able to elevate the dyke on its own, some 3,500 villagers working for 3 days placed 52,000 sandbags along the riverbank, repaired 69,000 m^3 of embankment and organised 24-h monitoring in different places (Municipal archive 2007) (see Fig. 4.8).

In disaster research some attribute this "disaster solidarity" to the practical benefits disaster victims receive (Hirshleifer 1988); others ascribe it to the process of coping with stress (Oliver-Smith 1999) or emphasise sacralisation of the action to optimise disaster response (Jencson 2001). However, the deep sense of community experienced in Ytyk Kyöl has been perceived by villagers as an amazing and positive aspect of the disastrous flooding. Another positive feeling the members of community have described is their ability to control their emotions and the absence of a panic atmosphere. Any talk about the flood event in 2007 in Tatta ultimately centres on

Fig. 4.8 Disaster solidarity: 'sandbag' community of Ytyk Kyöl (Photo: Alexander Postnikov)

recalling the tremendous sense of community produced during the "sandbag" operations. In addition to the shared hardship, this particularly heavy flood created literally bounded space as all roads were under water. The sense of commonality that arose spontaneously encapsulates for the politically "silent" Tatta communities their capacities for agency and a possible way of maintaining discretionary options.

4.8 "Processing" the Flood Experience

The attractiveness of the administrative centre in the Tatta District has increased the population density in what is a disaster-prone area. The infrastructure of Ytyk Kyöl expanded greatly in the last 30 years in the area very close to the water. During my first visit in 2008, a neighbourhood in direct proximity to the river was completely devastated and abandoned. The water in this area reached the top of the houses and remained there for almost 2 months. In the next year there were still abandoned houses but many of the homes in the area were already inhabited. People's "irrational and unfounded" return to areas devastated after a natural disaster, rebuilding in the same places and thus exposing themselves to the same risk have been documented in different regions of the world (Heijmans 2001; Leroy 2006; Berger 2008; Stammler-Gossmann 2010a) (See Figs. 4.9, 4.10).

Fig. 4.9 Yard during the flood in 2007 (Photo: Afanasii Lopatin)

Fig. 4.10 Yard after the flood in 2009 (Photo: Anna Stammler-Gossmann)

The aftermath of the flood has been dealt with less so far, as governmental assistance has focused on the disaster itself and reconstruction of damaged facilities. The compensation for those who suffered from the floods was a one-off payment only for persons directly affected and only for the duration of the flood (Decree N 542, 12.12.2008). Immediate measures, such as rebuilding the bridges in the area, were carried out, but some of the steps taken have caused unexpected consequences for residents. In the year after the 2007 disaster, a road embankment was built to protect houses, but only until a certain point, thereby blocking access to the houses beyond the embankment. The embankment has also disturbed the drainage system in the area and has kept the yards very moist, as I noted even 2 years later when I revisited Tatta in 2009. Gardening, an important supplementary activity in every household in the Tatta District, has suffered particularly severe impacts in this neighbourhood.

With no further financial incentives being allowed, residents have developed different practices to reduce their adaptation costs. Some of the houses whose owners were resettled or could not afford renovation are used as a summer cottages. In the houses where the people live permanently, basements were restructured and reinforced. New gardening practices were discovered and some households have experimented with new plants. As part of a school project on the impact of the flood on the potato yield, children conducted research on the possible treatment of infected vegetables (personal conversation).

Adaptive strategies are also the result of a process of innovation through which people build up not only their skills, but also the self-confidence necessary to shape their environment. As soon as the question of solving the drainage problem in the neighbourhood turned out to be a long-term problem, it was interpreted as an opportunity for individual action. After several appeals to the institutions that might be responsible for this matter, some of the residents dug drainage channels between their houses and the river. The administration made a few attempts to remove the channels as unsanctioned works, but finally accepted them. Individual actions brought more attention to consideration of the river regime in the rural infrastructure development and embankment quality, a question that arose in connection with the flood.

In my search during my fieldwork for the residents who were affected by flood events, many people independently suggested that I visit the house of a particular school teacher. Surprisingly, where I expected to see more damage, I entered one of the most beautiful houses in the village. Several sophisticated technical innovations in the house were introduced to me: a modified motorcycle to cross the inaccessible stretch from the embankment to the house, a platform in the house to move furniture in case of flooding, a device for lifting heavy household articles to the roof, and the like.

This personal engagement went beyond the actions of this single household and projected its impressive outcome in educational activities. The individual technical innovations were integrated into school lessons. A few new projects were initiated as an outcome of personal flood experiences, such as a questionnaire on the impact of the flood for Tatta residents, monitoring of changes in the phenology of the *alaas* meadows, gardening experiments, and vegetable storage practices impacted by the flood. It gave an impetus for broader activities in ecological education in the form of ecological training and summer school events. Knowledge of floods and practical advice on how to prepare for them were shared in the newly established regional newspaper and local TV programme. As one contributor to the newspaper stated, being prepared for a flood and its impact is more important than "different attempts to avoid and predict it; and preparedness depends mainly on people's own activity" (Lopatin 2010).

This view indicates that in a personal experience of agency the feeling of power to take action is an important factor in interaction with the natural and social world. However, surprise is frequently expressed that those whose homes and livelihoods have been destroyed by a disaster return to rebuild in rather than move out of the dangerous location. There are several reasons why this may be so. The building of houses close to the river was permitted, but people were also warned about exposure to possible floods. Nevertheless, villagers prioritised the affordances provided by living in proximity to the river: "You can fish, you always have water for the household and the garden and it is beautiful" (personal conversation 2009).

Gibson defines affordances as "action possibilities". They are latently offered by the environment, but always in relation to the actor and therefore dependent on his or her capabilities (Gibson 1979). Just as dynamic interaction with water is a part of everyday practice in Tatta, changes in the landscape of the plain are "internalised" in their autonomous flow and embedded into the process of acting. People's decisions have also been influenced by the perception that floods are rather unlikely and that there are "more advantages in living close to the river".

Merging human interactions with physical and social systems may bring further clarification to people's motives for living in a flood-prone location. As Yanitsky states, any risk under the totalitarian regime in the Soviet Union was treated as part of the indispensable price for the creation of a new society, as a price for the system's survival (Yanitsky 2000, 87). This high level of socially acceptable risk still remains in Russia and voluntary risk taking might still influence residential life in Tatta: "We knew about the risk, but there was a drought in the area that had lasted for several years".

4.9 Between Development and Natural Hazard

As was shown above, flood experiences have prompted local awareness of hazards resulting from water management projects.

The radicalised perception in Tatta villages has shifted the relations between "perceptible wealth and imperceptible risk", to quote Beck (1992, 44), to relations between advantages associated with "progress" and a risk that has become a numerical probability. It does not mean that the local perception of wealth is suppressed by the threat of a new flood.

Development projects still remain on the community's agenda. The drying up of the small Tatta River and its tributaries during the "dry" years may cause another form of disaster. The demand for water for irrigation remains high. The problem of providing a potable water supply is growing with increased pollution of water: Drinking water is taken from the river and lakes in ice blocks in winter and there are only very small artesian wells, whose water quality is poor. The flood has made these problems more acute and the community does not doubt the necessity of channelling water from the main pipeline to the village and the construction of dams.

The rhetoric of governmental plans, which emphasise a multiplicity of objectives in engineering construction projects, contributes to the ambiguity of the situation. The "Irrigation and water supply" project sounds very "social": Its "main purposes include restoration of the ecological balance in the region, providing clean and adequate water supplies for drinking and irrigation, flood control and fish resources enhancement" (Decree GS 25.04.2003).

There are no clear-cut solutions for how to balance both processes. Locals, who are so good in close monitoring of the river regime (e.g. flooding of grazing areas) and can deal with it in daily practice, may not be aware of behavioural responses that are required in the case of dams. As discussed previously, action in development projects is interpreted as a governmental responsibility. To the extent that they may impact irrigation constructions, floods are considered manageable only with state assistance. However, it is clear that the one-dimensional logic of wealth production at the expense of the environment increasingly conflicts with the growing concern of residents for protecting their environment as the foundation upon which development takes place.

Many people in the communities react ambivalently to the issue of the balance between "goods" and "bads". Hydro-technical achievements in the area were sharply criticised during my fieldwork in 2008, when signs of the disastrous flood of 2007 and of the smaller flood of 2008 were still visible. When I revisited Tatta in 2009, the perception of damage as a result of development projects had largely diminished: "We criticised our president for all these dams when our houses were under water but for this summer drought is predicted". The "manufactured" causes of floods and inadequate water management addressed in the official documents are never identified directly. Instead, administrative accounts focus on issues such as the technical condition of dams, achievements and failures of protective sanctions, historical accounts of dam breaks, calculation of possible impacts of regulated or unregulated water release, and local technical efforts to mitigate disaster (Municipal archive 2007).

"Managing ambiguities" generates ambiguity between ideas and action and produces specific agency. The perceived threats of the next possible flood that may be caused by water engineering are not always accompanied by direct actions. The extraordinary efforts on the part of the villagers in strengthening the earthen dam in Ytyk Kyöl during the flood had not been mobilised in conducting preventive and other countermeasures before and after the flood event. In-depth interviews revealed a profound ambivalence about what is believed and what is seen as the practical value of active engagement. The histories of changes in the physical environment are also histories of people and in this sense manufactured dams remain politically reflexive for them. Furthermore, it is estimated that the effectiveness of the community's or individuals' attempts to influence water management in the area through direct action would be low and not meaningful.

However, it would be wrong to talk about disengagement of Tatta residents when it comes to dam issues. Villagers do not struggle for support or oppose "rational strategies" but re-forge and re-address their concerns related to development projects. The question that the Ytyk Kyöl community is putting on the agenda today is not about natural versus technological – to build or not to build dams – but about "good" and "bad" dams. Recognition that natural and social processes go on in their own ways, a common view among residents, does not mean avoiding human interaction or influence on such processes.

Constructing a "good" dam in Ytyk Kyöl that can regulate water flow in the Tatta River and extensive recultivating of forest are nowadays among the main concerns related to flood protection. Bad experiences as downstream users of water have also prompted Tatta residents to ask about reconstructing earthen dams in the neighbouring district to include sluice gates. In this way, the concepts of "progress" or "prosperity" are reshaped into an awareness of a need for water management. Where the re-directed responsibility for dam issues to the state is more concerned with flood prevention, rural inhabitants shift their own responsibility to precautions to be taken in the event of floods. The municipal administration as an essential actor in the community's social network is seen also as a buffering institution between its ideas and deeds.

4.10 Conclusion

As the analysis shows, living with memories of flooding for Tatta communities means searching for a balance between natural processes and entities, self-organisation and regulations, and between risk and uncertainty. Prevailing institutional definitions of such extreme events as floods as departures from the norm are seen on the local level more as processes than as recurring patterns. While state activities in the area of adaptation are focused on first-tier agency intervention, local practices are related to sustaining livelihoods in the long term. During the flood in Tatta, the governmental frame described the risk of future floods in terms of economic damages, financial assistance, compensation and other "rationalities". However, uncertainties, which are perceived in this rural area as inherent in the natural and social environment, cannot be treated in that manner. People accommodate themselves to natural and social changes rather than eliminating those processes or entities. While the state risk approach is associated with "reworking" flood prevention and control, community agency is directed towards flood preparedness and regulation through "good" dams. Thus, the governmental and local agendas of "things to be governed" and "how they should be governed" may differ considerably.

Dealing with the ambiguity of meanings produces adaptive agency, where fragmented meanings can be accepted, accommodated, ignored, and resisted – sometimes, as the case study reveals, all at the same time. With local residents leaving themselves open to significant influence from the physical and social world around them, local decisions are shaped by interactions between components of the whole system. In this way, local adaptive mechanisms are often a starting point for adjusting to changes. At the same time, the mainstream community's agency obtains more efficiency by incorporating and entering into strategic "alliances" with institutional scales. Using informal bonds in interaction inside and outside of the state frame promotes more stability in the time of crisis and substitutes for the limits of state response. From this point of view, Tatta regional identity supported by considerable educational efforts may be seen as hampered by memories of marginalisation in the Soviet time but also as a force contributing to the building of a knowledgeable and responsible community prepared to invest in itself.

In their interaction, local people can perceive an environment that is meaningful for them and create a potential space for their agency. Following the view of affordances, or use-values, we can see the actions of Tatta residents as a process, as "the practitioner's way of knowing it", "discovering meanings in the environment through exploratory action" (Ingold 1992, 52–53). It means that a system that enables continuous learning and interaction between its components is more likely to facilitate innovative adaptation.

References

Adamov, D. (2010). Churapcha rulit! Nemnogo ob alasnom patriotizme (Churapcha governs! A little bit about *alaas* patriotism). *Yakutsk vechernii*, December 24.
ADRC (2006). Asian Disaster Reduction Center. Country report. Russia. http://www.adrc.asia/countryreport/RUS/2005/english.pdf. Retrieved 5 Dec 2009.
Alekseev, E. E. (1996). *Priznayu vinovnym… Sluzhba bezopasnosti respubliki Sakha (Yakutia)* (I declare you guilty of… Security Service of the Republic of Salkha (Yakutia). Moscow: Concern LD.
Alekseev, A. (2007). Nikto ne znaet tochno, v chem. prichina uchastivshikhsya navodnenii v Yakutii (Nobody knows exactly what the reason for increased floods in Yakutia is) *Sakha News*. http://www.1sn.ru/18234.html. Retrieved 7 Oct 2010.
APN (2005). Asian-Pacific Network for global change research. Institutional capacity in natural disaster risk reduction: A comparative analysis of institutions, national policies, and cooperative responses to flood in Russia. Final report. http://www.apn-gcr.org/newAPN/resources/projectBulletinOutputs/finalProjectReports/2005/APN2005-01-CMY-Nikitina_FinalReport-formatted.pdf. Retrieved 30 Nov 2009.
Beck, U. (1992). *Risk society, towards a new modernity*. London: Sage Publications.
Beck, U. (1999). *World risk society*. Malden: Blackwell Publishers.
Berger, A. R. (1998). Environmental change, geoindicators, and the autonomy of nature. *Geological Society of America today*, January, 3–8.
Berger, A. R. (2008). Rapid landscape changes, their causes, and how they affect human history and culture. *The Northern Review, 28*, 15–26.
Calhoun, C. (2004). A world of emergencies: Fear, intervention, and the limits of cosmopolitan order. *The Canadian Review of Sociology and Anthropology, 41*(4), 373–395.
EMERCOM (2004). Ministry of Russian Federation for civil defence, emergencies and elimination of consequences of natural disasters. Country report. http://www.adrc.asia/countryreport/RUS/RUSeng98/index.html. Retrieved 9 Apr 2010.
Forbes, B.C., & Stammler, F. (2009). Arctic climate change discourse: The contrasting politics of research agendas in the West and Russia. *Polar Research, 28*, 28–42
Gibson, J. J. (1977). The theory of affordance. In R. Shaw & J. Bransford (Eds.), *Perceiving, acting and knowing: Toward an ecological psychology* (pp. 67–82). Hillsdale: Erlbaum.
Gibson, J. J. (1979). *The ecological approach to visual perception*. Boston: Houghton Mifflin.
Heijmans, A. (2001). *'Vulnerability': A matter of perception*. Disaster management working paper 4. London: Benfield Greig Hazard Research Centre.

Hettinger, N. (2005). Respecting nature's autonomy in relationship with humanity. In T. Heyd (Ed.), *Recognizing the autonomy of nature: Theory and practice* (pp. 86–98). New York: Columbia University Press.

Heyd, T. (2007). *Encountering: Toward an environmental culture*. Aldershot: Ashgate.

Hirshleifer, J. (1988). *Economic behaviour in adversity*. Chicago: University of Chicago Press.

Ingold, T. (1992). Culture and the perception of the environment. In E. Croll & D. Parkin (Eds.), *Bush base: Forest farm. Culture, environment and development* (pp. 39–56). London/New York: Routledge.

Ingold, T. (2000). *The perception of the environment*. London: Routledge.

Jencson, L. (2001). Disastrous rites: Liminality and communitas in a flood crisis. *Anthropology and Humanism, 26*(1), 46–58.

Leroy, S. (2006). From natural hazard to environmental catastrophe: Past and present. *Quaternary International, 158*(1), 4–12.

Levochkin, V. (2008). Sezon vody (Season of water). *Rossiiskaya gazeta*, March 31.

Lopatin, A. (2010). Kihini erei yöreter (Disaster teaches human beings). *Sybehit, 1*, 2–3.

Lopatina, L. (2009). Puti sokraszeniia uszerba ot navodnenii na malykh rekakh (na primere reki Tatta). (Ways of mitigating the damage from floods on small rivers: the case of the Tatta River). In *Sbornik rabot molodykh issledovateli programmy'Shag v buduszee'* (Edited volume: Papers of young researchers in the programme: 'Step into the future', pp. 47–51). Yakutsk: Bargaryy.

Marková, I., & Gillespie, A. (Eds.). (2008). *Trust and distrust: Sociocultural perspectives*. Greenwich: Information Age Publishing.

Micklin, P. P. (1987). The status of the Soviet Union's north–south water transfer projects before their abandonment in 1985–1986. *Soviet Geography: Review and Translations, 20*, 287–329.

Oliver-Smith, A. (1999). What is a disaster? Anthropological perspectives on a persistent question. In S. Hoffman & A. Oliver-Smith (Eds.), *The angry earth: Disaster in anthropological perspective* (pp. 18–34). New York/London: Routledge.

Oliver-Smith, A. (2002). Theorizing disasters: Nature, power, and culture. In S. Hoffman & A. Oliver-Smith (Eds.), *Catastrophe and culture: The anthropology of disaster* (pp. 23–47). Oxford: School of American Research Press.

Regnum.ru (2004). Lenskaya voda prishla v Churapchu i Tattu (Yakutia) (Water of the Lena River came to Churapcha and to Tatta (Yakutia)). Regnum. http://www.regnum.ru/news/316970.html. Retrieved 7 Oct 2010.

Revich, B. (Ed.). (2008). *Climate change impact on public health in the Russian Arctic*. Moscow: United Nations in the Russian Federation.

Rose, N. (1996). The death of the social? Re-figuring the territory of government. *Economy and Society, 25*(3), 327–356.

Roshydromet. (2005). *Strategic prediction for the period of up to 2010–2015 of climate change expected in Russia and its impact on sectors of the Russian national economy*. Moscow: Roshydromet.

Salva, A. M. (1999). Cryogenic processes in engineering-geologic investigations for water-supply objects in transriverine regions of central Yakutia. *Hydrotechnical Construction, 33*(5), 307–312.

Stammler-Gossmann, A. (2010a). 'Translating' vulnerability at the community level: Case study from the Russian north. In G. K. Hovelsrud & B. Smit (Eds.), *Community adaptation and vulnerability in Arctic regions* (pp. 131–162). Dordrecht: Springer.

Stammler-Gossmann, A. (2010b). 'Political' animals of Sakha Yakutia. In F. Stammler & H. Takakura (Eds.), *Good to eat, good to live with: Nomads and animals in northern Eurasia and Africa* (pp. 153–175). Sendai: Tohoku University.

Stark, D. (2009). *The sense of dissonance. Accounts of worth in economic life*. Princeton/Woodstock: Princeton University Press.

Sukhoborov, V. (2006). Problems linked to the prevention of natural disasters in the Sakha Republic (Yakutia). The 3rd northern forum FWG meeting. http://www.yakutiatoday.com/events/inter_FWG_emercom.shtml. Retrieved 19 Sept 2009.

Vorobyev, D. (2005). Ruling rivers: Discussion on the river diversion project in the Soviet Union. In A. Rosenholm & S. Autio-Sarasmo (Eds.), *Understanding Russian nature: Representations, values and concepts* (pp. 177–205). Saarijärvi: Gummerus Printing.

WCDR (2005). National report of the Russian Federation at the world conference on disaster reduction. http://www.unisdr.org/2005/wcdr/preparatory-process/national-reports.htm. Retrieved 7 Dec 2009.

Yanitsky, O. N. (2000). *Russian greens in a risk society: A structural analysis*. Helsinki: Kikimora Publications.

Part III
Finnish Adaptation Governance

Chapter 5
Adaptation in Finnish Climate Governance

Monica Tennberg

Abstract This chapter presents and discusses the practices of Finnish climate governance with regard to adaptation. Finland takes its responsibility in climate politics seriously, although it is a rather recent idea that Finland could actually be vulnerable to direct impacts of climate change. Since the publication of the ACIA report (2004), a new governmental problematisation has emerged indicating that the people and nature in northern Finland could be "particularly sensitive" to the impacts of climate change. In its practices, Finnish adaptation governance mostly follows the logic of international climate governance, and it is expected to be even more internationally oriented in the future given the anticipated development of adaptation governance in the EU. It should be pointed out that half of the Finnish population is already covered by some kind of regional or local climate strategy that addresses adaptation to climate change. Adaptation strategies are now being drawn up for Lapland – the northernmost region of Finland – and for its capital, Rovaniemi. In Finnish adaptation governance, the role of the population is to be concerned, aware and willing to accept and participate in governmental action.

Keywords Climate change • Governance • Finland • Rovaniemi • Lapland • Adaptation

5.1 Finnish Climate Governance

Juhani Tirkkonen (2000, 135), an expert on Finnish climate politics, describes the development of Finnish climate governance as clearly linked to the development of international climate politics, in particular the Kyoto Protocol. Moreover, as Finland

---

M. Tennberg (✉)
Arctic Centre, University of Lapland, P.O. Box 122, 96101 Rovaniemi, Finland
e-mail: monica.tennberg@ulapland.fi

has been a member of the EU since 1995, the Union's climate politics have strongly influenced trends in the country; the EU has sought to become a global environmental actor in the issue area for the last decade. Much of the national debate has focused on the mitigation of greenhouse gases, because the country's northern location, sparse population and energy-intensive industries make its global share of greenhouse gas emissions relatively high. This means that it will be difficult for Finland to fulfil its international commitments to cut emissions, that is, to stabilise greenhouse gas emissions at the 1990 level by the year 2012 as a Kyoto commitment and to contribute to the EU's post-Kyoto commitment of a 20% emissions cut by 2020. Finland has been active in preparing for the impacts of climate change, adopting a national strategy for adaptation in 2005 and updating it in 2009. Adaptation in Finland is understood mostly as an environmental concern rather than a cross-cutting issue for the whole society (Jauhola 2010, 181).

This chapter presents and discusses the practices of Finnish climate governance with regard to adaptation. First, the problematisation of adaptation to the impacts of climate change in Finland is considered. Earlier national assessments suggested that the country might be vulnerable to indirect impacts such as global economic disturbances, conflicts and influxes of refugees. Recently, since the publication of the ACIA report (2004), a new problematisation has emerged indicating that the people and nature in northern Finland could be "particularly sensitive" to the impacts of climate change (Finnish Fourth National Communication under the UNFCCC 2006, 178). Secondly, Finland formulated a national strategy for adaptation in 2005 and evaluated it in 2009. More than half of the Finnish population is covered by some kind of regional or local climate strategy that addresses adaptation to climate change (Suomen Kuntaliitto 2009). Nevertheless, it is only recently that adaptation planning for climate change has started in northern Finland: Adaptation strategies are now being drawn up for Lapland – the northernmost region of Finland – and for its capital, Rovaniemi.

5.2 The Finnish Problem of Climate Change

The impacts of climate change on ecology, the economy and society Finnish have been studied since the mid-1990s in several research programmes and projects. The first such programme, SILMU (2010), which ran from 1990 to 1995, studied climate, water, territorial ecosystems and human activities. A second, "Finnish global change research", FIGARE 1999–2002, studied global change in the Finnish context (Käyhkö and Talve 2002). A research project called FINSKEN (Carter et al. 2004) developed projections of changes in environmental and related factors in Finland. In 2004–2005, a research consortium, FINADAPT (Carter 2007), led by the Finnish Environment Institute, studied adaptation to the potential impacts of climate change in Finland, including those seen as affecting biological diversity, forestry, agriculture, water resources, human health, transport, the built environment, energy infrastructure, tourism and recreation. The project also drew up a preparatory socio-economic study, urban planning measures, and a stakeholder questionnaire. Another project, the Climate Change Adaptation Research Programme, ISTO (2010), was started as part of the implementation of Finland's national strategy for adaptation to climate change in order to produce information for the planning of practical adaptation measures. The year 2009 saw the launch of two additional research projects related to vulnerability assessments, MAVERIC (2010) and VACCIA (2010), both coordinated by the Finnish Environment Institute. MAVERIC (2009–2012) is a map-based assessment of vulnerability to climate change employing regional indicators and VACCIA (2009–2011) is an assessment of the vulnerability of ecosystem services to climate change impacts and adaptation. A new research programme focusing on adaptation, FICCA (2010), operating under the Academy of Finland, has started recently. There are also a number of research projects in Northern Finland that focus on climate change and its impacts (Vihersalo 2008).

According to the final report of FINADAPT, a changing climate poses risks for the built environment, infrastructure and human well-being in Finland; yet it may bring some potential benefits as well. Where the national economy is concerned, the impacts of climate change on Finland will probably be rather modest and in the aggregate could even be slightly positive on average for the twenty-first century. This will be the case if climate change does not lead to abrupt, extreme events such as a disruption of the Gulf Stream. It is considered very difficult to assess the costs and benefits of climate change in economic terms. However, FINADAPT has estimated that that the order of magnitude of the aggregate effect will amount to less than a 0.1% change in the Finnish Gross Domestic Product. The economic sectors in Finland that could gain from climate change are forestry, agriculture and possibly also tourism. (Carter 2007, 60)

The first assessments of Finland's vulnerability to the impacts of climate change discussed the fate of the Gulf Stream and of major economic activities, especially forestry. In the mid-1990s, evaluations of vulnerability were hampered by the uncertainty related to the future behaviour of the Gulf Stream. There is uncertainty as to whether the Stream will lose its power or change its course, which would lower temperatures in Finland substantially. This could have a dramatic effect on the national economy, especially agriculture and forestry sectors. In the national 2001 report, the near-term concern for the Gulf Stream lessened following the completion of the Third IPCC assessment, according to which the probability of cooling was "negligible in this century, but it might increase thereafter" (Finland's Third National Report to UNFCCC 2001, 22).

National vulnerability assessments have highlighted the impacts of climate change on major economic activities such as forestry. For forestry, climate change would mainly mean a longer growing season and better conditions for forestry, although the concomitant increase in pests and diseases has generally been considered a threat. However, the expectation is that if the changes do not occur too rapidly, the forest industry will have time to adjust to them. All in all, the effects of climate change would seem to be less harmful in Finland than in many other regions. In some cases, Finland could benefit from climate change. For example, almost half of Finland's hydropower is generated in Lapland. Increased precipitation and more balanced discharges (smaller spring floods and larger discharges in winter) would be beneficial for hydropower production. (Finland's Fifth National Communication 2010)

The most severe problems Finland would face are the indirect effects of climate change elsewhere in the world. These could include influxes of refugees, lower food production due to desertification, and a general decline in the world economy. As one of the developed industrialised countries, Finland should contribute to prevention of such global problems and also participate in the costs of such preventive activities. According to Finland's 2001 national report, of all the risks the country faces due to climate change, those caused by detrimental changes elsewhere in the world will be the most serious. (Finland's national reports and communications 1994, 1997, 2001) The fifth national communication (2010) concludes that comparisons of the vulnerability of different countries to climate change indicate that Finland is among the least vulnerable.

It is only recently that the adaptation of people and livelihoods in northern Finland has been raised as a national issue. The Finnish fourth national communication (2006, 178) highlights the concern over the northern regions of the country: "Based on the present knowledge, the nature of northern Finland and its inhabitants will be particularly sensitive to the effects of climate change". This concern was raised by the ACIA report, published in 2004. The Finnish Arctic region has a population of some 180,000 people, of whom 4,000 are Sami. The climate changes observed thus far have been relatively small. However, Sami reindeer husbandry has been defined as a vulnerable livelihood. According to the assessment, reindeer husbandry is still an important economic activity in Lapland, particularly in small communities, owing to the meat it produces and the reindeer-based tourism it supports. Reindeer are also of great cultural value, because many of owners are Sami. (Finland's fourth national communication 2006, 167) In Finland's fifth national communication, submitted in 2010, the concern over the people and nature in Lapland broadened to include tourism. Tourism is the main industry in many communities in Lapland, and snow and the winter season are important for the industry.

Defining the problem becomes even more difficult from the domestic point of view, as the European Union has become more active in drawing up adaptation plans and strategies. The EU assessment is that climate change will heavily affect Europe's natural environment and nearly all sectors of society and the economy. In the 2007 EU Green paper on European Adaptation to Climate Change (EU Commission 2007) to climate change, both Scandinavia and the Arctic were defined as vulnerable areas. Scandinavia is considered vulnerable because much more precipitation is expected that will come in the form of rain rather than snow. The Arctic is deemed vulnerable, because temperature changes will be greater than in any other region on Earth. In the European Union context, the European Environment Agency (2005) report *Vulnerability and adaptation to climate change in Europe*, agriculture and forestry are the only sectors for which Finland has specified vulnerabilities (See also the Finnish response to the EU Green Paper; Maa- ja metsätalousvaliokunta 2008).

The development of the Finnish problematisation of climate change impacts and adaptation shows that science and politics are closely connected. The way issues and concerns become problematised seems to depend heavily on the amount of knowledge available and the role of the different governmental bodies that formulate issues and concerns in terms of vulnerabilities. For a long time, the Finnish national problematisation followed the international problematisation whereby climate change and its impacts are global problems. Only recently has it developed a problematisation of its own based on national and regional concerns, and some issues, such as climate change impacts on are still being discussed. However, this regional problematisation might be lost when the EU takes over the development of adaptation policies for the Union as a whole. Finland discovered its concern over climate change in its Arctic region relatively late. Locally, in Finnish Lapland, this concern is modest. The decision-makers in Lapland expect no major catastrophes due to climate change, but no major benefits either. The most positive impacts are expected in agriculture, energy production and forestry and with regard to infrastructure and transportation. The need for heating will be 12% less in the future compared to the average during the period 1970–2000 (Lapin liitto 2009a). Climate change also offers new opportunities for transportation and logistics in Lapland. The expectations relate to projected developments in the Barents region, which include a new maritime route from Europe to Asia (Lapin liitto 2009b). The most detrimental consequences are anticipated in the case of the local way of life and in reindeer herding, tourism and nature protection (Vihersalo 2007).

5.3 Adaptation Governance in Finland

In a period of some 10 years – from the mid-1990s to the mid-2000s – Finland discovered its need for national adaptation governance. In the mid-1990s, "given the uncertainty in the nature and magnitude of possible impacts of climate change", Finland did not see any need for the implementation of specific adaptation measures. By 2006, in its fourth national communication to the UNFCCC Secretariat, the country provided information on the expected impacts of climate change and on the projected adaptation measures as these were set out in the National Strategy for Adaptation to Climate Change published in January 2005. The need to draft a national programme for adaptation to climate change had been identified in the parliamentary debate on the National Climate Strategy submitted to Parliament in 2001. The work on the national adaptation strategy was coordinated by the Ministry of Agriculture and Forestry, with contributions from various ministries and expert organisations. The national adaptation strategy (Ministry of Agriculture and Forestry 2005) was evaluated and updated in 2009 (Ministry of Agriculture and Forestry 2009).

The objective of the strategy is to reinforce and increase the capacity of society to adapt to climate change. Adaptation may include minimising the adverse impacts of climate change as well as taking advantage of its benefits. The adaptation strategy extends to 2080. According to the strategy, climate change, its impacts and adaptation measures will take place over a very long time period. The relations between the impacts and cumulative effects are very complex and significant uncertainties are still involved. The Finnish adaptation strategy is not based on any estimation of the costs or benefits of adaptation measures. Finnish mitigation efforts and their costs are calculated and discussed in detail, whereas, in contrast, the monetary or other benefits of adaptation remain largely unassessed. Moreover, the strategy does not establish the relative priorities of different sectors with respect to climate change, although some sectors with high priority for the national economy have been identified.

According to the conclusions presented in FINADAPT (Carter 2007, 63), policymakers face the challenge of providing a policy environment within which autonomous adaptation can operate effectively. If policies are too rigid, then this "may stifle experimentation or entrepreneurial activity". In the worst case, ill-conceived policies may lead to mal-adaptation. The Finnish national strategy for adaptation is based on an assumption that "the private sector will benefit from the changes and will make autonomous adaptations to changing conditions".

Much of the adaptation required concerns public goods and public action. So far, most of the adaptation-related governance in Finland has focused on the work of the administration itself. The strategy is sectoral: All societal sectors will be required to assess and develop the facilities of the appropriate branch of administration, intensify the use of research information, and strengthen co-ordination and co-operation between different branches of administration, institutions and actors in order to promote adaptation. The means for adaptation in the strategy include administrative, legislative, economic and technical measures.

In addition to the national adaptation strategy, the Ministry of the Environment formulated, in 2008, its action plan on climate change adaptation, which included measures relating to biodiversity, land use and construction, environmental protection and the use of water resources. The Finnish national adaptation strategy of 2005 did not address national security as a sector in its own right. The government resolutions on the Internal Security Programme (Ministry of the Interior 2008) and the Strategy for Securing the Functions Vital to Society (Ministry of Defence 2006) deal with the preparations for climate change adaptation within the national defence administration.

Adaptation seems to be an issue quite independent of others in Finnish climate politics. For example, adaptation to the impacts of climate change is not included in the plans for mainstreaming Finnish climate governance. Mainstreaming simply refers to taking into account the concern over the climate in all governmental policies (Valtioneuvoston kanslia 2008a). Nor has adaptation been included in any of the national discussions on a Finnish Climate Act. The need for a special Climate Act has been discussed, but only with reference to emission cuts. According to a recent proposal for a Climate Act, the law would lead to cuts in national greenhouse gas emissions yearly by 5% leading to 38% cuts by 2020 and an 87% reduction by 2050 compared to the 1990 level. Corresponding laws are being drafted in the United Kingdom, Ireland, Scotland and Austria. (Valtioneuvoston kanslia 2008b)

An evaluation of the Finnish adaptation strategy (Ministry of Agriculture and Forestry 2009) was conducted in winter 2008–2009 through a survey of whether and how the measures presented in the strategy had been initiated in different sectors. According to the preliminary adaptation indicator developed in the context of this work, Finland, on average, is at step two in adaptation (on a scale from 1 to 5). This means that there is at least some understanding among decision-makers of the impacts of climate change and the need for adaptation measures. Some practical adaptation measures have also been identified, and plans have been made or even launched for their implementation. The sectors that have made the most progress in implementing the adaptation strategy have been water resources management, the transport sector, community planning, agriculture and forestry. However, in most sectors the work is "only in early stages". The implementation of adaptation measures also calls for more co-operation between sectors, especially at the regional level.

The Finnish evaluation of the national adaptation strategy (Ministry of Agriculture and Forestry 2009) estimated that future adaptation policies and strategies will become more European than at present. The rationale for a European approach lies in the need to "ensure proper co-ordination and the efficiency of policies that address the impacts of climate change" among the EU member states (See also EU Green Paper 2007). According to the latest EU plan, set out in a White Paper (EU Commission 2009), the EU aims to have an EU-wide adaptation strategy starting from 2013. Jauhola (2010, 169) estimates that the role of the EU will be in adaptation governance in terms of "enabling adaptation fostering networks within the wider regions and building capacity in the absence of direct national support".

5.4 Adaptation Governance in Finnish Lapland

It was not until 2009 that local and regional adaptation work began in Finnish Lapland. Overall, many Finnish municipalities participate in the International Climate Network of Cities (ICLEI). According to a recent study by the Association of Finnish Local and Regional Authorities (Suomen Kuntaliitto 2009), 26 municipalities in Finland have made their own climate strategies or programmes, either by themselves or in collaboration with other municipalities. Such plans are now being made in an additional 54 municipalities. All in all, one-fourth of Finnish municipalities are engaged in active climate politics. This activity covers 64% of the Finnish population. In the other municipalities, concern over the climate has been noted in other municipal plans and strategies.

Locally, for example in the city of Rovaniemi in Lapland, climate change adaptation is defined and determined by nationally set laws, regulations and programmes. In addition, municipalities can make climate policies and strategies of their own that aim to save energy and develop infrastructure, for example. In Rovaniemi, the plans for a city climate programme (not a strategy) were started in 2009. A draft city programme (Rovaniemen kaupunki 2009a, b) states in a very general way that climate change will pose some challenges for the city and that they will respond to them. Local concerns are tourism, energy production and waste management. One important tool in adaptation is city planning, especially preparedness for flooding of the Kemijoki River, which runs through the city centre. According to a recent study (Järviluoma and Suopajärvi 2009), the case of Rovaniemi shows that adaptation to the impacts of climate change requires plans at not only the community, but also the regional and national levels. However, among the participants in recent local and regional climate strategy making there has been some confusion as to how the city plans and regional strategies are connected, or if they are at all.

Preparations for climate change have also started at the regional level and include a year-and-a-half long project to develop a regional climate strategy for Lapland that was started in spring 2010. This is an activity of the Regional Council of Lapland, which with support from EU funding through the Centre of Trade, Environment and Transportation, has hired consultants to take care of the strategy-making process. The main consulting company is Bionova, which has several subcontractors to collect background material, such as the Meteorological Institute and University of Lapland, and to carry out the strategy-making process, such as Capful and Eurofacts. The aim of the strategy is to provide a plan for mitigating and adapting to climate change regionally until 2030. The strategy envisions Lapland in 2030 as "a successful region that takes both opportunities and challenges of climate change into consideration".

The formulation of a regional climate strategy for Lapland follows the logic of Finnish climate governance in general: It is mostly administrative, sectoral and expert-based, and most likely provides limited opportunities for public discussion and debate (See Lapin liitto 2011). Participants in the strategy-making process come from various regional administrative bodies, research institutes, major companies and various interest organisations. Despite the effort to engage participants and organisations broadly from different parts of Lapland, participation by representatives from different communities and region in Lapland has been limited to those with a special interest in the issue. These include delegates from the Kemi-Tornio region, which has strong industrial interests, the City of Rovaniemi, which has undertaken its own climate programme work, or Sodankylä, which is pursuing regional economic interests. Participation of the Sami – the local indigenous people – or other non-governmental organisations has been modest in the process.

The strategy work has focused on defining the challenges and opportunities presented to Lapland by climate change. The strategy involves mitigation and adaptation mainly in various nature-based activities, such as forestry, reindeer herding and fishing, tourism, and the development of mining. The regional emissions of greenhouse gases from a national perspective are modest and in a general sense Lapland is a carbon reservoir because of its large forest areas. Much of the debate has been embedded in a discourse of opportunities, such as the business opportunities in opening northern sea routes, new technologies to save energy, and various kinds of innovations. In terms of adaptation, many challenges relate to changing winter conditions, the delayed onset of winter, variation of temperatures and increasing precipitation in winter time. The fate of the tourism industry is a major economic issue.

On the other hand, Lapland has emerged as an alternative to ski resorts in central Europe which have suffered from a lack of snow in the beginning of the winter season; it expected that Lapland will become a more popular destination for alpine skiing. But even in Finnish Lapland, the delay of winter has caused some problems for the tourism industry. The lack of snow in mid-November, when the Christmas season starts, requires developing new products for tourists that are not as snow-based and snow-dependent. Moreover, as tourists become more environmentally aware, their interest in how the tourism business takes environmental issues into consideration grows. In Lapland, alternatives to emission-producing air traffic are particularly important for tourists. In addition, there is a need to develop services in hotels and resorts to make them more environmentally conscious, for example, by increasing opportunities for recycling and better waste management (See also Lapin liitto 2007).

Several elements of the regional climate strategy have already been discussed in other regional strategies, such as in the regional energy strategy (Lapin liitto 2009a), the strategy for the tourism industry (Lapin liitto 2007), and the regional plan for 2030 (Lapin liitto 2009b). Adaptation to climate change and its impacts at the regional level in Lapland seems to have its own "realities" vis-à-vis concerns at the national level. The national strategy and its recommendations do not seem to figure in the regional discussion. A European-wide study on national adaptation plans has noted that the links between the plans at the local and regional levels and the Finnish national strategy are weak or lacking, although one can find several references to actions to be taken at the local and regional levels (Swart et al. 2009).

The urban planning study (Carter 2007, 68) carried out as part of the FINADAPT project identified some obstacles to an awareness of climate change concerns among local planners. These were lack of training, a perception of climate change as a long-term problem, the marginality of the issue compared to other immediate problems, and ambiguities at the ministerial level between mitigation and adaptation in the guidance given to planners. According to the results of the FINADAPT study, the issue of adaptation should be taken up at all levels of spatial planning. In this way, awareness would be raised and co-operation between various actors in planning improved. In addition, regional, integrated studies of climate change impacts and adaptation are needed in preference to the conventional sectoral studies. The sectoral approach to planning could produce some institutional hindrances to adaptation governance.

The experience of Lapland so far confirms some of these ideas: The most important aspect of regional climate strategy-making is not the strategy paper itself but the process. This means involving participants from various parts of and sectors in Lapland and enabling them to discuss with each other and share knowledge and views about climate change, its impacts and the possible adaptation needed. The implementation of regional strategy and the process that follows it some years after – an assessment and a review of the impacts of the strategy – will show the success of the regional climate strategy. In the ideal case, no new regional climate strategy will be needed; rather, the regional problematisation of climate change and its meaning is "internalised" and implemented in various kinds of action plans and strategies made by the regional administration, companies and organisations and concrete and practical measures to mitigate and adapt to climate change will follow. As the regional strategy process has suggested climate change "should be a part of the everyday life of decision-makers and inhabitants".

5.5 Aware But Politically Inactive Citizens

In the development of the national adaptation strategy in the mid-2000s, the Finnish stakeholders had an opportunity via the Internet to participate in the formulation of the national strategy for adaptation. Among the Finnish population, climate change is a real concern for the majority: Nine out of ten Finns consider it a major threat to humankind. In comparison to other problems, climate change was ranked third in importance in 2002 and 2005; the more urgent concerns were economic recession, drug use and terrorism. In 2007, one-third of the respondents identified climate change as the most important threat to them. Storms, floods and rains – all closely related to climate change – were identified as threatening. In 2007, only 5% of respondents envisaged that climate change and its impacts would not have an effect on their lives. Women consider the potential impacts more serious than men. Comparisons have shown that Finns are somewhat more concerned about climate change than other Europeans. (Valtioneuvoston kanslia 2008c, 15–17)

The majority of the Finnish population think that climate change is a human-induced problem. Citizens considered it a threat that has been very poorly prepared for and regard international measures to tackle the problem as being important. The majority also feel that Finnish politicians should take climate change more seriously than they have thus far. Furthermore, most think that adaptation to the impacts of climate change must be started immediately. (Valtioneuvoston kanslia 2008c, 20) Half of the respondents in the FINADAPT stakeholder study thought that the measures taken so far in Finland to adapt to climate change have not been sufficient: One-fourth considered the actions adequate. In general, Finns are willing to take actions to tackle climate change, such as recycling, saving energy and using public transportation. They have a strong belief in their own willingness, but a lack of trust in the willingness of others to do the same. (Valtioneuvoston kanslia 2008c, 25)

Awareness of climate change among the Finnish population is high but people lack opportunities to influence and participate in climate politics. A recent study on Finnish climate co-operation by Kauppila and Savikko (2009) defined the main "climate" actors as research institutes, some ministries, environmental and consumer organisations, municipalities, companies, political decision-makers, and expert activists. "New" climate actors include regional councils, women's organisations, the Church, advertising companies, immigrant organisations, national defence, and private homeowners. The research showed that citizens, civil society, and participation in climate work mean very different things to different participants.

According to Kauppila and Savikko research, Finland was seen as a country with a relatively flat social hierarchy and a consensus-oriented culture. The positive side of this is direct contacts with decision-makers; one drawback is a lack of debate. Significant expert knowledge and expertise are valued by all categories of agency. Participation in climate work is based on administrative leadership and in-group work with only a limited number of people who have suitable backgrounds, knowledge and expertise invited to take part in this work. The research concludes that there is no broad, empowering societal debate on international climate co-operation and Finnish commitments. In current practice, participation and influence are seen as parallel, as if the participation as such of an environmental organisation in a working group or delegation is enough. According to environmental organisations, this is not the case. Moreover, problems and the solutions to them are discussed within the limits established by the government. The researchers conclude that in Finland the general belief is that politics are the responsibility of politicians and that one can best influence issues by influencing politicians. The Finnish consensus-oriented political culture does not respond easily to the demands for broad participation with multiple concerns and voices.

Vihersalo (2007) studied the awareness of climate concerns among decision-makers in Finnish Lapland. A questionnaire was sent to 266 respondents, of whom 130 replied. The respondents considered the climate a genuine concern, one with already observable consequences. Climate change is considered to be a problem that can be controlled, but only to a slight extent. In general, the respondents were somewhat concerned about climate change. The problem was seen as caused by both human and natural factors. The researcher considers it important that respondents viewed climate change as a natural phenomenon, which makes it a difficult problem to manage. This is a challenge for the regional climate strategy: how to make local decision-makers and inhabitants take climate change seriously? In the regional climate strategy process, the inhabitants of Lapland are expected to become even more aware of climate change and its impacts, to become more ethically motivated consumers and more energy efficient in all aspects of their everyday life, including housing, work and leisure.

In our own analysis of the regional newspaper *Lapin Kansa* (2002–2008) (Sinevaara-Niskanen 2009), we identified three different framing strategies by which local citizens are made responsible, knowledgeable and involved in climate change politics. The research suggests that people are willing to be governed as part of adaptation governance. In the first frame, citizenship, which includes not only rights but also responsibilities as citizens, people are presented with some normative expectations of the individual and collective action required to mitigate and adapt to climate change. The newspaper articles indicate that citizens are expected to act because "environmental protection is based on human choices". These choices are controlled and directed through instructions, incentives and sanctions ("carrots and sticks will be used"). "Ordinary citizens are woken up", and they are expected to pressure politicians to make things happen. This "responsible citizen", or "responsibilisation", frame balances between a citizen's role and a trust in individual choice and freedom. The frame is based on the attitudes and actions expected of a good citizen.

The second local frame embraces a "knowledgeable and normal" actor. The associated texts describe how households and individuals should act, drawing on knowledge as the basis of that guidance: Research results, reports and experts describe the right ways – the "normal ways" – of doing things and making choices. Expressions that support what "normal people" do, such as "the majority of population", "almost 90%", or "three times more", support this "normalising" frame. Knowledge-based framings give readers factual information to both support and help them in decision-making. When "two thirds of Finns save energy" or "when 70% change their lamps to energy lamps", there is no reason why the rest should not follow their lead. Such knowledge-based frames construct worries and raise awareness. The knowledge that is produced and offered should be taken seriously and, at the same time, the concerns of the majority should produce more concern and normalise it. As a result, it is not normal to be not concerned. This "normalising" frame creates a population that is more aware and better able to maintain concerns, even globally.

Third, there is a frame that produces involvement, or participation. Whereas the frames of "responsibilisation" and "normalisation" base the relationship between individual and community on citizenship, guidance and knowledge, the "participation" frame stresses autonomous action. It is based on "us" talk, the importance of choices and alternatives, and the presence of individuals ("I at least would be ready to…", "When I understood things, I started to think…"). Participation and the choice between alternatives are connected to economic arguments. The use of energy and choices of means of transportation are supported by economic reasoning as reflected in statements such as "it makes economic sense to modernise your heating system". The question of making choices is not based on ecological awareness but directed by economic conditions and interests. Making one's own choices requires one to "internalise" a concern and to grow into a certain kind of an actor, that is, an economically rational actor.

5.6 Conclusions

Finland takes its responsibility in international climate governance seriously. Finnish adaptation governance is an exercise in the top-down, expert-based and administrative management of society. Much of its development can be explained by participation in international and Arctic climate co-operation and its influences on Finnish governance. The Finnish view is that the main threat to the country lies in the indirect impacts of climate change, such as global economic disturbances, conflicts and influx of refugees. The planning of adaptation started early in comparison to many other countries. Finland discovered its concern over climate change and its impacts in the country's Arctic relatively late. Since the publication of the ACIA (2004), a new governmental problematisation has emerged indicating that the people and nature in northern Finland could be sensible to the impacts of climate change (Finland's fourth national communication 2006, 178). The particular concerns in the region are reindeer husbandry and the tourism industry and their capacity to adapt to changing environmental conditions.

In terms of governance practices, Finnish adaptation governance mostly follows the logic of international climate governance, and in the future it is expected to be even more internationally oriented due to the development of the EU's adaptation governance. However, at the same time, half of the Finnish population is now covered by some kind of regional or local climate strategy that takes into account adaptation to climate change (Suomen Kuntaliitto 2009). It is only recently that planning for adaptation to climate change has started in northern Finland: Plans are in the making for Lapland – the northernmost region of Finland – and for its capital, Rovaniemi. The regional strategy follows the national logic: top-down, expert-based and sectoral administrative planning for the future. The national concern over the fate of reindeer herding in the future due to harmful impacts of climate change does not seem to translate into any special measures to promote adaptation regionally. Due to the sectoral approach in Finnish adaptation governance, problems arise at the regional level in the form of weak or lacking linkages between sectors and the administrative levels of activities (See also Peltonen 2009).

In Finnish adaptation governance, the role of the population is to be concerned, aware and willing to accept and participate in governmental action. Knowledge plays an important part in defining the role of the citizen. Knowledge about climate change and its impacts plays a central role in the "responsibilisation", "normalisation" and "participation" of citizens in governance. The scope of a citizen's actions is mostly economic: Individuals are expected to make wise economic choices in their everyday lives. The traditional scope of political action and participation seems limited, confined to representation by various NGOs in the official structures of decision-making. It is noteworthy that people are amenable to being governed and to being part of adaptation governance.

References

ACIA (2004). Arctic climate impact assessment. http://www.acia.uaf.edu/pages/overview.html. Retrieved 16 June 2010.

Carter, T. R. (2007). *Assessing the adaptive capacity of the Finnish environment and society under a changing climate: FINADAPT. Summary for policy makers*. Finnish Environment 1/2007. http://www.environment.fi/default.asp?contentid=227529&lan=en&clan=en. Retrieved 21 June 2010.

Carter, T. R., Fronzek, S., Bärlund, I. (2004). FINSKEN: A framework for developing consistent global change scenarios for Finland in the 21st century. *Boreal Environment Research. 9*, 91–107. http://www.borenv.net/BER/pdfs/ber9/ber9-091.pdf. Retrieved 21 June 2010.

European Commission (2007). *Adapting to climate change in Europe – options for EU action*. Green paper COM(2007) 354 final. http://eur-lex.europa.eu/LexUriServ/site/en/com/2007/com2007_0354en01.pdf. Retrieved 21 June 2010.

European Commission (2009). *Adapting to climate change: Towards a European framework for action*. White paper. COM(2009) 147 final. http://eur-lex.europa.eu/LexUriServ/LexUriServ.do?uri=COM:2009:0147:FIN:EN:PDF. Retrieved 25 Aug 2010.

European Environment Agency (2005). *Vulnerability and adaptation to climate change in Europe*. EEA Technical Report 7/2005. http://www.eea.europa.eu/publications/technical_report_2005_1207_144937. Retrieved 25 Aug 2010.

FICCA (2010). Research programme on climate change FICCA. http://www.aka.fi/en-gb/A/ Science-in-society/Research-programmes/Open-for-Application/Climate_change-ficca/. Retrieved 25 Aug 2010.
Finland's fifth national communication under the United Nations framework convention on climate change (2010). http://unfccc.int/resource/docs/natc/finnc5.pdf. Retrieved 22 June 2010.
Finland's fourth national communication under the United Nations convention on climate change (2006). http://unfccc.int/resource/docs/natc/finnc4.pdf. Retrieved 19 June 2008.
Finland's national report under the United Nations framework convention on climate change (1994). http://unfccc.int/resource/docs/natc/finnc1.pdf. Retrieved 19 June 2008.
Finland's second national report under the United Nations framework convention on climate change (1997). http://unfccc.int/resource/docs/natc/finnc2.pdf. Retrieved 19 June 2008.
Finland's third national report under the United Nations convention on climate change (2001). http://unfccc.int/resource/docs/natc/finncd.pdf. Retrieved 19 June 2008.
ISTO (2010). Climate change adaptation research. http://www.mmm.fi/en/index/frontpage/ymparisto/ilmastopolitiikka/researchprogrammeonadaptationtoclimatechange.html. Retrieved 21 June 2010.
Järviluoma, J., & Suopajärvi, L. (2009). *Ilmastonmuutoksen ennakoituihin vaikutuksiin sopeutuminen Rovaniemellä.* CLIMATIC-hankkeen raportti (Adaptation to climate change in Rovaniemi. A report by the CLIMATIC project). Lapin yliopiston yhteiskuntatieteellisiä julkaisua C. Työpapereita 52. http://www.ulapland.fi/loader.aspx?id=c2935fbf-c199-4de7-9a34-4966379adce4. Retrieved 29 June 2010.
Jauhola, S. (2010). Mainstreaming climate change adaptation: The case of multilevel governance in Finland. In C. Keskitalo (Ed.), *Developing adaptation policy and practice in Europe: Multilevel governance of climate change* (pp. 149–188). Berlin: Springer.
Kauppila, J., & Savikko, R. (2009).*"Kyl se sit niissä verkostoissa tapahtuu se päätöksenteko"* – *Kansalaisvaikuttaminen ilmastopoliittiseen päätöksentekoon Suomessa* ("It is those networks where the decisions are made". Public participation in climate politics in Finland). http://www. globalplatform.fi/files/kansalaisvaikuttaminen_ilmastopoliittiseen_paatoksentekoon....pdf. Retrieved 21 June 2010.
Käyhkö, J., & Talve, L. (2002). Understanding the global system – the Finnish perspective. FIGARE 1999–2002. http://www.sci.utu.fi/projects/maantiede/figare/UGS/UGS.pdf. Retrieved 21 June 2010.
Lapin liitto (2007). *Lapin matkailustrategia 2007–2010* (Lapland's strategy for tourism 2007–2010). http://www.lapinliitto.fi/c/document_library/get_file?folderId=53864&name=DLFE-3 210.pdf. Retrieved 29 June 2010.
Lapin liitto (2009a). *Lapin energiastrategia* (Lapland's energy strategy). http://www.lapinliitto.fi/c/ document_library/get_file?folderId=53864&name=DLFE-3202.pdf. Retrieved 29 Oct 2010.
Lapin liitto (2009b). *Lappi – Pohjoisen luova menestyjä. Lapin maakuntasuunnitelma 2030* (Lapland's regional plan 2030). http://www.lapinliitto.fi/c/document_library/get_file?folderId =53982&name=DLFE-3226.pdf. Retrieved 29 June 2010.
Lapin liitto (2011). *Lapin ilmastostrategia 2030* (Lapland's climate strategy 2030). http://www. lapinliitto.fi/fi/lapin_kehittaminen/strategiat/lapin_ilamastostrategia. Retrieved 15 Aug 2011.
Maa- ja metsätalousvaliokunta (2008). *Valtioneuvoston selvitys komission vihreästä kirjasta: sopeutuminen Euroopassa – vaihtoehdot EU:n toimille* (Governmental report on the EU commission's Green Paper: Adaptation in Europe – options for Europe). Lausunto 372008 vp. http://www.eduskunta.fi/valtiopäiväasiakirjat.htm. Retrieved 14 Jan 2009.
MAVERIC (2010). Map-based assessment of vulnerability to climate change employing regional indicators MAVERIC. http://www.environment.fi/default.asp?contentid=318210&lan=EN. Retrieved 25 Aug 2010.
Ministry of Agriculture and Forestry (2005). Finland's national adaptation strategy to climate change. http://www.mmm.fi/attachments/mmm/julkaisut/julkaisusarja/5kghLfzOd/ MMMjulkaisu2005_1a.pdf. Retrieved 21 June 2010.
Ministry of Agriculture and Forestry (2009). Evaluation of the implementation of Finland's national strategy to climate change. http://www.mmm.fi/attachments/mmm/julkaisut/ julkaisusarja/2009/5IEsngZYQ/Adaptation_Strategy_evaluation.pdf. Retrieved 21 June 2010.

Ministry of Defence (2006). *Strategy for securing the functions vital to society (Yett strategy).* Government resolution 23.11.2006. http://www.defmin.fi/files/858/06_12_12_YETTS__in_english.pdf. Retrieved 25 Aug 2010.
Ministry of the Environment (2008). *Adaptation to climate change in the administrative sector of the Ministry of the Environment.* YMrep20en/2008. http://www.ymparisto.fi/download.asp?contentid=90891&lan=fi Retrieved 15 Aug 2011.
Ministry of the Interior (2008). *Safety first – internal security programme.* Government resolution. Ministry of the Interior publications 25/2008. http://www.intermin.fi/intermin/biblio.nsf/B48B12C5D837461AC22574C00025B90A/$file/252008.pdf. Retrieved 25 Aug 2010.
Peltonen, L. (2009). Aluetason toimintamallit hakemassa muotoaan (Regional models for adaptation take shape). *Kuntatekniikka, 3*, 28–31.
Rovaniemen kaupunki (2009a). *Kaupunkistrategian luonnos 2009.* Draft for a city strategy. http://www.rovaniemi.fi/loader.aspx?id=f6a59dc7-289f-422d-8918-51f0a950ee06.Retrieved 29 June 2010.
Rovaniemen kaupunki (2009b). *Kaavoitusohjelma 2009–2012.* Perusteluosa (luonnos) (City planning program 2009–2012, draft). http://www.rovaniemi.fi/loader.aspx?id=cddd3f48-4963-4f18-8846-8df0296df40c. Retrieved 29 June 2010.
SILMU (2010). *Suomalainen ilmakehän muutosten tutkimusohjelma* (Finnish research program on climate change). http://www.aka.fi/fi/A/Tiedeyhteiskunnassa/Tutkimusohjelmat/Paattyneet/Suomalainen-ilmakehanmuutosten-tutkimusohjelma-SILMU-1990-95/. Retrieved 25 Aug 2010.
Sinevaara-Niskanen, H. (2009). Memo: Analysis of *Lapin Kansa* newspaper articles 2002–2008.
Suomen Kuntaliitto (2009). *Kuntien ilmastonsuojelukampanja* (Campaign for climate protection in municipalities). http://www.kunnat.net/k_peruslistasivu.asp?path=1;29;356;1033;36689;36692. Retrieved 25 Aug 2010.
Swart, R., et al. (2009). *Europe adapts to climate change. Comparing national adaptation strategies.* PEER report 1. Vammala: Saastamalan kirjapaino.
Tirkkonen, J. (2000). *Ilmastopolitiikka ja ekologinen modernisaatio. Diskursiivinen tarkastelu suomalaisesta ilmastopolitiikasta ja sen yhteydestä metsäsektorin muutokseen* (Climate politics and ecological modernisation. Discursive study of Finnish climate politics and change in the forest sector). Acta Universitatis Tamperensis 781. Tampere: Tampereen yliopistopaino.
VACCIA (2010). Vulnerability assessment of ecosystem services for climate change impacts and adaptation. http://thule.oulu.fi/vaccia/. Retrieved 25 Aug 2010.
Valtioneuvoston kanslia (2008a). *Ilmastopolitiikan valtavirtaistaminen ja politiikkakoherenssi. Selvitys Vanhasen II hallituksen tulevaisuusselontekoa varten.* (Mainstreaming climate politics).Valtioneuvoston kanslian julkaisusarja 6. http://www.vnk.fi/julkaisukansio/2008/j06-ilmastopolitiikan-valtavirtaistaminen/pdf/fi.pdf. Retrieved 25 Aug 2010.
Valtioneuvoston kanslia (2008b) *Selvitys Iso-Britannian ilmastolakiehdotuksesta ja alustava arvio vastaavan sääntelyn soveltuvuudesta Suomen oikeusjärjestelmään* (Report on the climate bill in the United Kingdom and a preliminary analysis of the applicability of comparable legislation in the Finnish legal system). Valtioneuvoston kanslian julkaisusarja 16. http://www.vnk.fi/julkaisukansio/2008/j16-selvitys-ison-britannian-ilmastolakiehdotuksesta/pdf/fi.pdf. Retrieved 13 June 2011.
Valtioneuvoston kanslia (2008c). *Ilmastoasenteiden muutos ja muuttajat.* Selvitys Vanhasen II hallituksen tulevaisuusselontekoa varten (Changes in and variables affecting attitudes on the climate). Valtioneuvoston kanslian julkaisuja 9. http://www.vnk.fi/julkaisut/julkaisusarja/julkaisu/fi.jsp?oid=237072. Retrieved 13 June 2011.
Vihersalo, M. (2007). Ilmastomuutos Lapin päättäjien näkemyksissä. ZEF-kyselytavan käyttö moniulotteisen ympäristöongelman tutkimisessa (Decision-makers' views on climate change in Lapland). In S. Ronkainen (Ed.), *Ilmastonmuutoksen ennakointi ja vaikutusten arviointi Lapissa. Raportti sähköisen kyselyn ja kaksiulotteisen kysymisen toimivuudesta globaalimuutosta tutkittaessa.* Lapin yliopiston menetelmätieteiden julkaisuja 2. http://www.ulapland.fi/loader.aspx?id=8f1c62d7-0823-4a24-98c5-1c34b8dccbc6. Retrieved 10 June 2010.
Vihersalo, M. (2008). *Ilmastonmuutosta koskevat strategiat ja raportit sekä ilmastonmuutosta käsittelevien hankkeiden kartoitus* (Strategies, reports and projects on climate change). http://thule.oulu.fi/oyly/Ilmastonmuutoshankeselvitys_Vihersalo.pdf. Retrieved 10 June 2011.

Chapter 6
Adaptation of Sámi Reindeer Herding: EU Regulation and Climate Change

Terhi Vuojala-Magga

Abstract Northern Sámi reindeer herders have adapted to various changes throughout their history. One source of the flexibility and resilience that these changes have required is the traditional Sámi self-governance system of *siidas*, which is based on kinship among reindeer herding families. The most recent challenges facing herding are connected to climate change and EU regulations on reindeer meat production and carnivore protection. This chapter views climate change not only as a threat, but also as a possibility in light of the new regulations. The *siida* system, which features a range of members from different generations, creates a dynamic interaction that generates new, skilful innovative solutions to offset constraints on the livelihood: for example, in the case of reindeer meat production, people have managed well in spite of new hygiene regulations and climate change. In contrast, the effect of regulations protecting carnivores has been that the *siida* system and people's skills no longer offer adaptive solutions. This could be the first time in the history of Sámi reindeer husbandry that the younger generation does not see any future in their livelihood.

Keywords Reindeer herding • Sámi • Climate change • Adaptation • EU • Carnivore management

6.1 Introduction

Reindeer husbandry in Finnish Sápmi, the Sámi homeland, has faced a variety pressures and changes in the course of the twentieth century. The history of Sámi reindeer husbandry has shown that it has been one of the most powerful factors

T. Vuojala-Magga (✉)
Arctic Centre, University of Lapland, 122, 96101 Rovaniemi, Finland
e-mail: vuojala-magga@suomi24.fi

M. Tennberg (ed.), *Governing the Uncertain: Adaptation and Climate in Russia and Finland*, DOI 10.1007/978-94-007-3843-0_6,
© Springer Science+Business Media B.V. 2012

contributing to the cultural survival of indigenous Sámi societies during times of hardship. However, two new challenges have emerged recently: the European Union, with its regulations, and climate change. The recent EU regulations on meat production have led to new practices and expanded local entrepreneurship, and EU directives on predator hunting pose new challenges to reindeer husbandry. This chapter discusses the EU regulations on meat production and carnivore protection in the context of climate change. The question of adaptation is approached through the concepts of agency and governance and their basic dynamics for coping with changes coming from outside of reindeer husbandry. The focal agent is the individual reindeer herder, and governance is defined in terms of the *siida, a self-created* organisation of reindeer herders and their extended families. Of principal interest in the discussion here are the *siida* system and its dynamics in the face of the changes discussed.

6.2 Reindeer Herders' Views on Climate Change

Climate change is often seen as a phenomenon somewhere beyond the range of human beings and their activities. As Tim Ingold points out: "/../ Rather than the (global) environment surrounding us, it is we who have surrounded it" (Ingold 2000, 154, 215). However, we all are a part of climate change. Changes in life-worlds caused by the warming climate are plainly visible among northern indigenous peoples, because their economies rely directly on nature and are subject to the impacts of climate change on the texture of snow and ice. Sámi reindeer herders are part of climate change as agents – among others – in the development of meat markets and policies regarding predators. Contemporary studies on reindeer herding and climate change have brought the discussion of governance to the fore as well. The main problem seems to be that reindeer herders' traditional ecological knowledge is somehow ignored in the decision-making processes (Tyler et al. 2007; Eira et al. 2009; Forbes and Stammler 2009; Beach and Stammler 2006; Eálat home pages 2010; Lehtola 1997; Mazzullo 2010; Tuomas-Aslak Juuso 2010; Juha Magga 2010; Baer 2010; Rees et al. 2008).

This chapter focuses on questions of agency and governance from the perspective of the reindeer herder. Of particular interest are the herder's seasonal tasks – meat production and struggles with predators – within the dynamics of the *siida* system and amid efforts to adapt to changes affecting the livelihood. The emphasis is on generational changes in the kinship system underpinning the *siida* in the context of EU regulation and climate change. The study is based on ethnological research focusing on one *siida* and its members – the Kuttura village *siida*[1] *of the Hammastunturi* herding co-operative, which is located on the upper course of the Ivalo River in the municipality of Inari in Sápmi (see Fig. 6.1). In Inari, the

---

[1] The village of Kuttura village is located near the border of two reindeer herding co-operatives, Hammastunturi and Sallivaara, and is therefore divided into two siidas.

Fig. 6.1 Aerial view of part of the village of Kuttura (Photo: Urho Minkkinen)

Sámi people are members of one of three language groups: the Inari Sámi, Northern Sámi or Skolt Sámi. Some of the herders in the area are Finnish. The Kuttura *siida* discussed here comprises five households and a total of some 20 members, including children. They all belong to the Magga family, who are Northern Sámi. They earn their living either partly or entirely from reindeer herding.

The local livelihood, social settings and adaptation to changes are viewed in terms of the practical engagement of individuals. Practice in this sense refers to both practical mastery of reindeer herding skills and the dynamics of the local social systems – in this case, the *siida* – as these are experienced by individuals (See Ingold 2000, 162). Knowing, thinking and understanding are generated in the practical tasks and work of everyday life (Lave 1990, 310, 1991, 14; Lave and Wenger 2001). Practice is seamlessly connected to the concept of activity and links individuals to the lived reality (Kozulin 1996, 112–113). Because activities (or actions) and bodies are 'constituted' within practices, 'the skilled body' is where activity and mind, as well as individual and society, meet (Schatzki et al. 2001, 3). Tim Ingold has argued that human beings and animals are not passive objects of their life courses, nor are they only genetically programmed organisms or objects of cultural transmission. Rather, we should focus on active persons, described by Pallson as "the whole persons in action, acting within the context of that activity" (Pallson 1991, 904). Hence, the concept of agent as it pertains to the northern environment refers to an innovative and active person within that environment. As Ingold states: "We do not have to think the world in order to live in it, but we do have to live in the world in order to think it" (2000, 418). In the same vein, in the present case, instead of talk *about reindeer husbandry there is a talk within the practice* (see also Lave and Wenger 2001, 120)

In this chapter, the perspective is that of the agent, a reindeer herder, in his or her social context, the *siida*. The local perspective defines adaptation governance as a possibility to negotiate and act within a group of reindeer herders and as action in these circumstances in response to the changes taking place within reindeer husbandry. The essential question is whether adaptation governance is based on "practices by which political rationales are *translated* to practical actions" (See Chap. 1) or on practices and actions *in continuous transformation* governed by agents and their communities.

6.3 The *siida*: A System of Sami Self-governance in Reindeer Husbandry

From the point of view of reindeer herders, the organisations associated with reindeer husbandry form a two-fold system: the first tier comprises a ground-level organisation of actors who are originally from reindeer-herding communities or are their representatives; the second consists of faceless or not personally known authorities who manage the framework in which actions are taken and regulations and laws are implemented.

The basic system of reindeer husbandry in the Sámi area is the *siida*, which consists of a Sámi reindeer village comprising several families or one family who work together on a daily basis. A *siida* can vary and function in different ways according to the area and type of reindeer herding involved. It is defined as a unified system composed of economically independent households. A *siida's* success in herding depends on its degree of consensus and the members' ability to act and exchange information in accordance with the knowledge and insight gained by regular participation in the daily life of the system (Oskal 1995; Joks 2000; Sara 2001; see also Paine 1994).

A *siida* defined in terms of practices is an open system, changing over time and in accordance with socio-economic needs. In Finland, a *siida* comprises a single extended family or a certain number of households tied together by kinship. Each member has his or her personal tasks geared to furthering the success of the reindeer herd and the members of the *siida*. In this respect, the number of members in each *siida* can vary yearly: during climatically good years, *siidas* may be small, and during bad years larger. During a time of personal conflicts, a *siida* can be small and during a time of good personal co-operation large. The *siida* system is rather flexible.

Decision-making processes in *siidas* vary according to the tasks at hand. Where collective work is involved, such as collecting reindeer from the forest or organising round-ups, decisions are negotiated by the reindeer herders working in the forest. However, the most powerful reindeer herder, that is, the one with the largest number of reindeer, has the last word. Most often he or she is one of the oldest and the most experienced owners and the head of one of the families in the *siida*. The number of reindeer of this highly regarded person is seen as a whole; it is the number which the household/family members own, not only the number with that individual's particular earmark. Ultimately, power is not concentrated in the hands of a single person but in the hands of a family of active herders, who then authorise that individual to make final decisions for the *siida*. The philosophy of power pertaining to big owners is not so much one based on their being rich and thus powerful; rather, I would speak of a "major owner" as being one who has achieved success in reindeer husbandry. However, success is not only a matter of good fortune and circumstances; in its deepest meaning, it is about sound skills and hard work in herding among highly motivated and inspired members of the younger generation of the same family.

The *siida* system varies from one herding co-operative to another. Although it is based on kinship, it is flexible in accommodating the herding techniques of each co-operative. Young, active single men outside of the main family may be hired to work in *siidas* and may eventually become "external members" who then invigorate the younger generation of the *siida*. Power relationships in a *siida* can also change if the younger generation of a powerful family is not active in reindeer herding work; it can steer other families of active reindeer herders in a different direction, because the future of the livelihood is always based on the number of active younger members of the next generation. Everyone knows that there is a future to consider, too, and the power relationships are viewed in the long term. In this respect, the *siida* is a highly social system and the dynamics of changes in its power relationships prompt actions and drive its internal dynamics.

In addition to the Kuttura *siida* comprising the Magga family, there are three other villages in the Hammastunturi herding co-operative: Menesjärvi (the Jomppanen, West and Mattus families), Inari (the Nikula, Lehtola, Kangasniemi and Huovinen families) and Huuhkaja (the Huhtamella and Magga families). A village can have one or several *siidas* depending on the size of the village and the families in it. The villages and their *siida* system are so strong in the Hammastunturi co-operative that reindeer have a pasture rotation throughout the year in the area of each village and *siida*. This means that autumn and winter round-ups are also organised by each *siida* and its members around their and their reindeer's home area.

Siidas are part of the national system of reindeer herding co-operatives. There are 56 such co-operatives in northern Finland, of which 13 are situated in Sápmi. Finnish authorities created the management structure of reindeer herding on the basis of this eastern tradition of reindeer husbandry at the end of nineteenth century (Reindeer Herding Decree 1898, 1916; Reindeer Herding Act 1932; Kortesalmi 2003, 2007, 382–385). Each co-operative is organised by it its members, who know each other very well since they work together on numerous occasions. The election of the head of a herding co-operative (poroisäntä) is carried out using a system based on the numbers of reindeer, with each herder getting as many votes as he or she has animals[2] (Reindeer Herding). This could be called democracy based on the number of reindeer, which in turn reflects success in herding and the *siida*. In addi-

---

[2] However, no one is allowed to cast a number of votes that exceeds 5% of the total number of reindeer owned by the partners in the co-operative.

tion to receiving national funding, a co-operative gets an annuity from the reindeer owners, known as "head money", for each reindeer. Each reindeer owner has to pay the annuity to the co-operative either as a sum of money based on the number of reindeer or by doing a certain number of working days. The head of a co-operative ensures compliance with national laws and collective decisions; he or she has to see to it that all the day-to-day work is done and thus assigns tasks to the herders. Heads of co-operatives have to be impartial with regard to the work that is done. Moreover, they should have the full confidence of the members of the co-operative, as they represent it in different outside organisations, negotiate on its behalf in difficult situations and work with the media. Lehtola and Mazullo comment that the co-operative system featuring a head owner is considered part of the Finnish but not the Sami system: "The head of a co-operative has nothing to do with the *siida*; the co-operatives give power to one person, who does not obey the regulations and traditions of the *siida* system" (Mazzullo 2010). In this same vein, Juha Magga, in a speech given in 2010, pointed out that *siidas* have no legal rights to take part in different organisational meetings (Magga 2010).

However, on the level of reindeer herding practice, the head of the co-operative is elected according to the power relationships within the *siida* and their interaction. In some cases, it can be extremely hard to make impartial decisions, for example, in dividing tasks between herders if there are more workers than work. In some cases, the head can be more or less a head on paper only, with certain skilful members of the co-operative giving the final orders. Yet, there have to be some other benefits to electing a head: he or she is chosen to meet various other needs in the co-operative, such as the need for a person with excellent skills in negotiating with other *siidas* or with other organisations and the media.

6.4 Changing Practices of Reindeer Meat Production and the *siida*

When Finland joined the EU in 1995, reindeer husbandry became subject to EU governance. From the point of view of reindeer herding and management, the most significant changes brought by EU legislation have been the regulations on reindeer meat production and how animals are to be slaughtered. Compliance with these provisions has been monitored and implemented by the national organisation of the Ministry of Forestry and Agriculture (MAF), the Reindeer Herders Association (RHA), and the Centres for Economic Development, Transport and the Environment (ELY) (Kortesalmi 2007, 399). One of the aims is the authorities is to guide individual reindeer herders and co-operatives to conduct their own meat production in keeping with the European regulations. In Finland, new, small slaughterhouses and processing facilities were established in various co-operatives using EU money as well as national funds. By the end of 2005, there were 40 slaughterhouses and processing plants operating in accordance with EU regulations (Kortesalmi 2007, 399). In another aspect of the modernisation process, a number of hygiene rules

Fig. 6.2 EU-compliant reindeer slaughter house in Ailigas (Photo: Terhi Vuojala-Magga)

came into force in 2006 (Euroopan parlamentin ja neuvoston asetukset EPNA Ey 2004; MMM. muistio 26.6.2009; Meristö et al. 2004; Raito and Heikkinen 2003; Heikkinen 2002, 2006; Saarni et al. 2007; Tauriainen 2009; Tuomisto 2008: Tuomisto and Jauhianen 2008).

However, all modernisation takes time and old practices for slaughtering animals and processing and selling meat continued alongside the new. Even though EU programmes supported new investments geared to improving local entrepreneurship for individual herders, those taking advantage of the investment programmes were mostly young herders. Investments were made by individuals (the West family in Solojärvi, part of the Hammastunturi co-operative), *siida*s as a unit (the Kustula families in the Ivalojoki co-operative), and entire co-operatives, such as Kaldoaivi in the Utsjoki region. The modern system of meat processing and production can be understood as a response to the needs of urban customers and, at the same time, as a step towards traditional European entrepreneurship, which is based on liberal individualism. Hygiene regulations, in turn, are based on central European meat production practices, developed for the densely populated and climatically mild areas of Europe (see Fig. 6.2).

Small individual entrepreneurship is nothing very new among reindeer herders; it was already widely practiced before the 1960s. In the old days, reindeer roundups were held a few times per year. Meat buyers used to come to the round-up fence and the meat price was negotiated between each reindeer herder and buyer. Small

Fig. 6.3 Old open- air slaughter site by the Sotajoki enclosure (Photo: Terhi Vuojala-Magga)

reindeer herders, that is, those who had relatively few animals, would join forces with owners who had larger herds. They sold their reindeer together, bringing the smaller owner a better price per kilogram for live animals than he or she would have obtained otherwise (Male 1950 a). In most cases, this type of co-operation took place among the members of the same *siida*. Where meat was sold directly to a buyer, the reindeer were herded live to slaughtering places or were slaughtered in an open slaughter yard by the round-up fence (see Fig. 6.3).

A number of technological innovations were introduced in reindeer husbandry after the 1960s: snowmobiles replaced skis and the road network improved. Snowmobiles made Sámi herders' life easier in that they no longer had to be in the forest for weeks and months and instead of having a few huge round-ups annually herders began to arrange them more regularly. In the case of big companies, meat prices were negotiated beforehand, and all the reindeer owners of the same co-operative sold live reindeer to the companies at the same price. At the same time, the system for negotiating meat prices and sales changed from one based on individuals and *siidas* to one based on co-operatives. Until 1995, companies such as Lapin Liha and Rönkkö were the big buyers and live animals were transported by truck to Kemi and Rovaniemi (See also Heikkinen 2002, 2006). In another development, consumption started to favour calf meat after the 1960s. As a result, what had been mixed herds with both breeding and castrated males changed to become

Fig. 6.4 Reindeer leg skins for fur shoes (Photo: Terhi Vuojala-Magga)

homogenous herds of productive females. Male calves were slaughtered, which made it possible to increase the size of herds.[3]

Already after the 1960s, the system by which meat was sold became more diverse: traditional slaughter and sale at the round-up fence were not abandoned but became transformed into a market serving private individuals, members of the reindeer herding family and local needs. Slaughter and direct sale enabled herders to keep up their traditional skills in handling carcasses without extra waste; that is, blood, intestines, bones and leg skins were used with optimal efficiency. Traditionally, tenderloin and sirloin are not separated from the spine, because herders prefer meat soup made from tenderloin and sirloin, as well as blood sausages and blood cakes. The skins are also used locally and this gives a good basis for continuing traditional Sámi handicrafts and a good opportunity for artisans to make products from reindeer fur and leather (see Fig. 6.4). With the help of the national organisations RAF and MAF, reindeer herders obtained a special dispensation from the EU for their traditions because of the highly hygienic manner in which the work is carried out.

Today, there are many ways indeed of slaughtering reindeer and processing, producing and selling the meat (see Figs. 6.5 and 6.6). This obviously benefits reindeer herders and customers, because it opens up a broad range of different

---

[3] Twenty per cent of a herd should consist of male reindeer.

Fig. 6.5 A private reindeer meat processing facility and shop, Tundra Poro. Reindeer herder and entrepreneur Visa Valle in front of his shop (Photo: Terhi Vuojala-Magga)

Fig. 6.6 Reindeer meat processed by Visa Valle (Photo: Terhi Vuojala-Magga)

markets and alternatives. As the slaughter is carried out by reindeer herders, who eat a lot of reindeer meat, slaughtering, cutting and transportation are all handled appropriately. As one herder commented: "The meat sold is like a business card; it has to be perfect, because it is a matter of our personal prestige" (Male 1948). However, the new procedures also pose new problems. First, those with small companies suffer from a serious lack of time, as there is far too much work during the round-up, slaughter, and meat processing season. Second, the profitability of reindeer herding seems to favour owners of large herds; small owners cannot invest in and adjust to modernisation of the meat market as readily (see also Heikkinen 2006).

6.5 Changing Practices in the Kuttura *siida*

In Finland, each co-operative has different ways of herding reindeer. The reindeer in the Hammastunturi co-operative graze around each *siida*. Because of this, each *siida* decides what is the most suitable time to gather reindeer for the autumn round-ups, which again are arranged near each *siida*.[4] The Kuttura herding district is situated in the area of the continental divide, where the landscape features small rivers, canyons and pine-spruce forests; it is known as an area with deep snow. The round-up site for the Kuttura *siida* is the Sotajoki enclosure. It is an area where reindeer can only be gathered for autumn round-ups once rivers have a permanent ice cover and there is enough snow to enable herders to catch animals hiding in deep canyons. Once the seasonal conditions for work are optimal, the herders of the Kuttura *siida* are ready to collect their reindeer. The final decision for the start of a round-up is given by the head of *siida*, the most skillfull member and the head of the most powerful family. The herders work in pairs and after negotiations with the head and his or her family members, decisions are made on the strategy for collecting the reindeer. Normally it takes from 1 to 2 weeks to gather the reindeer into the enclosures, with the work completed by the middle or end of October. The ultimate driver of herding in the Kuttura *siida* is the coming of the right season and weather – ample cold and snow.

However, ice and snow conditions changed during the 2000s; autumns are now warmer and longer and have no snow and ice. In recent years, round-ups have not been held until the second week of December. As a result, the weight of the reindeer is lower than before, decreasing herders' income. Herders have estimated that a calf loses from 1 to 2 kg per month from November on, which translates into a loss of income per 100 calves per month of at least 900–2700e or more depending on the

---

[4] For comparison, the three co-operatives of Sallivaara, Kaldoaivi and Paistunturi have only a single round-up site, at which all of the reindeer are gathered.

grade of meat. In mid-December, the losses of the Kuttura *siida* totalled between 1200e and 4050e per 100 calves.[5]

Today the members of the Kuttura *siida* have a variety of means for producing reindeer meat, part of a long-term trend towards more diverse forms of production. Until the early 2000s, most of the reindeer were sold to meat companies as before, and the traditions of outdoor slaughter by the round-up fences continued in each family. However, individual decisions have been made tacitly by each owner and his or her family. The private meat business is improving year by year. The year 2009 marked the first time that there was only one reindeer herder who wanted to sell his reindeer live to the large company Lapin Liha. The most influential families sold their reindeer live to the local Ivalo co-operative company and the others carried out their slaughtering themselves. In this case, the one reindeer herder had to make a decision in keeping with the practices of the others, because the meat company's truck would not be coming north again and would not come to collect a single family's few reindeer. Eventually the herder had to sell the family's reindeer along with those of the most influential family of the *siida* to the Ivalo co-operative. Some families slaughter their reindeer privately at the slaughterhouse of the Toivoniemi Reindeer Research Station, where the carcasses get a stamp from the veterinarian. The meat is cut and packed at home in hygienic storerooms and eventually transported by each reindeer herder and sold privately to customers and intermediaries in the south.

In 2010, the younger generation of the *siida* members – those born in the 1960s – became frustrated and managed to hire a helicopter to help in collecting reindeer immediately after the rut. This was the first time that a reindeer round-up was held at the beginning of October, a time when there is no snow or ice. The herders were very successful. They drove the reindeer into the enclosure in 3 days and the number of animals was higher than usual. Slaughter and sale were completed by the end of October. This new, successful way of organising round-ups indicates that the old structure and power relationships of the Kuttura *siida* are changing: the younger generation is taking over and in a few years' time there will be a new head of the *siida*.

6.6 EU Carnivore Management Policy: A New Problem for Reindeer Herders

The way in which meat is sold and the market for meat are changing gradually, without any deep crises. Although reindeer herding is a co-operative form of work, the different types of individual sale systems and neo-liberalism seem to function

---

[5] For comparison, Paistunturi combines the herds of different siidas, and the first round up is arranged after the rut at the end of September, with animals gathered for slaughter every fourth or fifth day thereafter. By December most of the reindeer in the co-operative have been sold, and thus the herders benefit from long, warm autumns.

smoothly among herders, although there were some worries at the very beginning of 1995. However, at the moment, the worries caused by the EU come from its carnivore management policy (Habitats Directive). The number of predators has risen to a level that threatens the entire livelihood in parts of the reindeer herding regions. Juha Magga has pointed out the problem: "It is a fact that the number of animals falling prey to predators is increasing all the time. In parts of the co-operatives most of the income is disappearing into the mouths of carnivores; the percentage of calves is only 30–40%" (the number of calves born per 100 female reindeer) (Magga 2010; see also Kavakka 2009; Magga 2009).

The number of reindeer found killed by predators has increased radically during Finland's membership in the EU. Before the country joined the Union in 1995, there were fewer than 1500 reindeer killed by predators in the reindeer herding area, whereas in 2007 the number found had risen to 4,090 (Paliskuntain yhdistys 2008, 29; Lapin liitto 2011, 8). However, figures never tell the whole story: most of the reindeer killed are never found. Carnivores are spreading throughout the reindeer herding region, with the situation being most serious in the southernmost and eastern regions. The number of calves slaughtered has dropped radically in the southernmost co-operatives: the decrease was 23% and the calving percentage was only 25%. The same tendency is spreading north-eastwards and into central Lapland (Norberg 2010, 19; Danell and Norberg 2010).[6]

Administrators on the national level are aware of the problem, as are the organisations of reindeer herders. The new head of the RHA has made carnivore policy the main topic of discussions between different organisations in the near future.[7] By 2010 there were few proposed solutions to the problem, e.g. an increase in the compensation paid for lost or killed reindeer and the enactment of harsher the criminal sanctions for illegal predator hunting. However, from the perspective of reindeer herders, these actions would not offer a constructive solution. Instead, the herders' focus is on the realities and possibilities of practicing reindeer herding for a living. In the southern Halla co-operative, the threshold has been reached at which "artificial respiration" is taking place; it is no more a question of compensation. The final attempt of the southernmost regions to survive can already be seen in the system by which reindeer meat is sold and compensation paid: Reindeer herders from the south come to the northernmost reindeer herding districts, for example the Paistunturi co-operative, to buy live reindeer in order to maintain their livelihood. Paradoxically, compensation for killed reindeer is transferred to the north and live reindeer are

---

[6] The most common carnivore in northern Finland is the brown bear. Between 1978 and 2007, the bear population increased from 300 to 880–950 individuals, with the number of animals increasing to 1,050–1,300 during the summers. The population of lynx (*Lynx lynx*) increased between 1978 and 2009 from 100 to 1,400 individuals. The number of wolves (*Canis lupus*) rose between 1978 and 2007 from 80 to 340 and the wolverine population (*Gulo gulo*) increased from 50 to 80 individuals in 1980 to some 155–170 in 2007. There is no permission to hunt wolverines. (MMM 6/1996; Kainulainen 2008; RKTL 2007).

[7] A forum for discussion on carnivores was held in Salla 31.1.2011.

Fig. 6.7 Female reindeer killed by a wolf (Photo: Visa Valle)

brought to the south, with the meat market ultimately losing in terms of the number of reindeer. In these encounters and personal contacts between southernmost and northernmost reindeer herders, anxiety and worrying news spread. The distress is palpable and real, and makes everyone think about his or her future in the livelihood (see Fig. 6.7). At the moment, the most common feeling among reindeer herders is fear. As one said: "We have this kind of permanent feeling of distress; it affects every aspect of our lives" (Male 1950b).

As noted earlier, the reindeer of Kuttura live in an area with deep snow. In the late winter months of March and April, the reindeer eat arctic moss on pine and spruce trees. During those months the snow cover should be hard due to sunny and warm days combined with clear and cold nights; such weather makes it easy for the forest reindeer to graze from tree to tree (Vuojala-Magga et al. 2011). However, lately there have been years when the crust on the snow has not been hard enough, or has been made up of or many thin layers, known in Finnish as *sevä*. This is rather difficult for full-grown reindeer, for they cannot move properly in the snow. Only the lighter calves are able to graze easily.

Carnivores such as lynx and wolverines benefit from these changing snow conditions. When an adult reindeer has difficulty running in deep snow with thin layers of ice, injuring its legs, it becomes easy prey for a wolverine. The snow has melted earlier than before during the past 5–6 years (Vuojala-Magga et al. 2011). For bears, early-melting snow affords excellent conditions for hunting: they benefit from open patches of ground. As a reindeer herder said in criticising officials: "There are airplanes patrolling day after day in the air and we know it is expensive to patrol every single day. We just wonder where all these airplanes and people on snowmobiles are once a local gets lost in the forest; they only come after a few days. Is the monitoring of bear movements more important than searching for a lost person?" (Reindeer herder male 1950a).[8] The hardest time for reindeer is during the months of May and June. The new-born calves – known as "milk meat" or "milk calves" – are no doubt the most delicious food for bears. Small calves get tired running long distances, making them easy prey. Reindeer move in herds, and once there is an easy catch, predators get greedy and kill more than they need. As one reindeer herder commented: "Predators are like human beings: once there is plentiful food on the table, they tend to eat the best parts of it" (Reindeer herder 1950b).

The calving percentage in the Hammastunturi co-operative was rather low during the 2000s. Hunger or weakness is not the reason for the low number of calves; the reindeer have been in good condition. The decade was climatically constant, with longer autumns and shorter springs. The short winter provides reindeer with easy and longer access to lichen and other ground flora. Reindeer herders have adapted to the soft snow conditions, and during difficult springs they bring hay into the forest to keep the reindeer well fed (Vuojala-Magga et al. 2011).

In *siidas*, the earmark system reflects a means for compensating herders who have lost reindeer to predators. If one of the active members of the family with relatively few reindeer faces the threshold of viability with his or her herd, new calves born to the reindeer of a richer family member (e.g. a big owner) can be marked in the coming year using the earmark of the less fortunate herder.[9] However, the total number of reindeer sold decreases for this family, which means a lower income. This same logic is being applied between co-operatives as well, with live reindeer being sold from north to south. Losses of reindeer are reflected in the declining numbers of young herders: when there are no more calves – a new generation – there will be no more young generations of herders. The fundamental issue here is the future of reindeer husbandry from the perspective of the younger generation.

The agent – in this case, the reindeer herder – is limbless; illegal killing of predators is no answer to the problem. Intensive herding is being practiced in most of the northern co-operatives, with each herder tending his or her reindeer and herds being constantly driven from one area to another during the spring months. However, in forest areas such as the pasture area of the Kuttura *siida* reindeer cannot be kept in one herd, as they spread out through the forest in spring to feed on arctic moss.

---

[8] In one case where a young local man disappeared in the forest, the air patrols did not start to look for him for 4 days.

[9] The discussion of everyone in the EU having the rights to engage in reindeer herding in Finland sounds like pressure on the representatives of Sámi reindeer herders. In practice, obtaining an earmark in a Sámi co-operative is a lengthy process. If an earmark is being sold in a co-operative, another herder from the same co-operative has precedence in buying the animals. An owner without a family and a *siida* will receive no support if he or she suffers losses.

In the case of a carnivore attack, permission to kill the predator comes as quickly as possible, that is, in a few days' time, but as a rule the reindeer that are killed and eaten are not found as often as one would hope.

In reindeer husbandry, herders have learned to live with and accept the hard years with extreme climatic conditions or predator killings; by nature has upswings and downswings reindeer husbandry. However, hardship has traditionally referred to temporary, not lasting difficulties. Collapses have happened in almost every decade, but people have always recovered from losses. Predators have never been viewed as a problem or as a reason to abandon one's livelihood, for herders have had the possibility of controlling the situation. In the old days, the attitude of reindeer herders towards predators was one marked by a certain "respect and control." In one way, the skilful action of the animals prompted a respect among people and in the case of severe losses the skilled action of the reindeer herders was applied to control the situation.

On the administrative level, discussions and negotiations regarding the problems are under way and the compensation systems have improved.[10] Until 2009, compensation for a reindeer killed by a predator was paid only if the animal was found and was in the forest. However, in 2009, a new Act on damage caused by game animals was passed which provided that compensation is paid according to the decline in the percentage of calves and the estimated percentage of calf deaths between May and November (Kavakka 2009, 6). This is an improvement, as most of the calves killed are never found. A second improvement is that compensation for exceptionally large losses, where a reindeer is found killed, is paid at three times the value of the animal (Kavakka 2009, 6). Yet, the policy differs markedly from that in the other Scandinavian countries; for example, according to Kojola the rate of compensation in Norway is eight to nine times higher (Kojola 2009). A different system is applied in the case of losses caused by eagles (Ollila 2009).

EU policy seems to be bringing herding to the brink of collapse as the reindeer population declines. Compensation can be seen as a corrective process for the time being, but as predators spread northward there are no more possibilities of transferring reindeer from one area to another and thus no means of avoiding future losses. People used to say: "We never know all the deaths traps for calves – nature gives and nature takes" (Male 1950a). In the old days, the concept of nature included human beings and their actions in the forest. Today, the management of predators and reindeer is no longer a balancing of good and bad years, but merely the control of outside forces.

6.7 Agency and Governance in the *siida* System

Climate change as an ill-structured problem (See Chap. 7, Hartwell paper 2010) can be analysed by using the North as a construction representing a life-world of agents and their social systems within their environment of everyday practice. In reality,

---

[10] The latest research done by Lapinliitto was published in 2011. It looks as if the predator problem is spreading towards the co-operatives of the far north. (Kainulainen 2011)

the northern environment and reindeer herding itself are bound by a social system of practical activities and skills: no one survives alone. In Sámi societies, an agent is somehow always linked to his or her family – and in this case the *siida*. As a kinship system, the *siida* can be seen as a unit of self-governance. Its basic dynamics and generative potential lie in its ever-changing power relationships based on generations and individual skills that contribute to success in reindeer herding.

An agent can be defined through his or her activity. Being an agent in a *siida* means being present: agents are fully absorbed in its work. Innovations or old ways of doing things in herding are negotiated or objected to in discussions, but ultimately everything is tested in practice on a daily basis. This is also described as "understanding in practice". In this context, enskilment plays an important role, meaning that individuals as a part of the community express meanings based on their practical knowledge – be it of the present situation of climate change, meat markets or predator policy. Reindeer herders' knowledge relates directly to their capacity for acting in light of past, current or future situations and actions.

A *siida* and its governance system are never stable, but have a dynamic based on yearly success. The older generation has its time and the new generation makes its presence felt through its own activities and systems (see also Lave and Wenger 2001). These dynamics of co-operation, innovation and the power relationships among individuals, families and generations shape the *siida* system to make it perfectly functional. The *siida* as a unit of a social system uses different individual skills for the common good and, at the same time, individuals can apply their skills to further their personal success in reindeer herding. Hence, co-operation and competition exist in parallel, and success and hardship individual provide actors with various ways of demonstrating resilience and adaptive capacity. However, a *siida* is more than individual actors carrying on their present and future tasks, for family units form microsystems within a *siida*. The dynamics of the power relationships within the *siida* diminish if there is no younger generation planning ahead, competing and bringing in new ways of doing things.

In the case of meat processing, EU regulations have prompted individual agents, families or even co-operatives to find new ways of coping. Agents – individual reindeer herders – have various possibilities to choose the best and the most suitable way to process the meat they produce for market. Neo-liberalism is nothing new on the individual level. Even though the new EU regulations sounded dire in their early stages, there has been a great deal of flexibility in development on the local level. The changes have taken place in 15 years' time, with herders observing the experiences of different co-operatives. However, in the case of investments, an agent is never alone; a new entrepreneur will have the backing of his or her entire extended reindeer-herding family. Here the system differs from that seen among Finnish herders, which is based on agriculture and in which the basic unit is a single household. There might be a close family in the background, but hardly ever the extended family that functions as a system for exchanging help and support in the way that a *siida* does (see Fig. 6.8).

As the markets for selling meat have improved, many reindeer herders have expanded their business and are competing with big companies in buying more

Fig. 6.8 Reindeer skins on sale in front of a small cafeteria and a shop run by the Nuorgam family in the village of Kaamasmukka (Photo: Terhi Vuojala-Magga)

meat for private sale from their reindeer herder colleagues. Ultimately, these successful practical actions have served as a message to many other herders to shift to meat processing, private sale and marketing. Instead of viewing traditionalism as a stable system in the case of meat processing by Sámi herders, the concept can be seen as a chain of changes over decades. Even though the pressures for changes in meat production have come from the outside, they have been positive ones if analysed in terms of productivity and the income of individual agents and their families. The success of an agent depends on his and her enskilment in reindeer herding, co-operation and the governance of the *siida*.

Climate change has affected reindeer herding systems in the Kuttura *siida*. The long autumns in the 2000s postponed round-ups for months, and the *siida* system, with its tradition of a head herder, has not been reacting to these changes. Since 2009, there has been quiet pressure from the younger generation for changes, and in 2010 the round-ups were arranged earlier than before by the younger generation with the help of new technology. Historically, the *climate can be seen as a governor of people*: they accept it as they have accepted different weather patterns for centuries. Herders have adopted and will adopt different techniques and take a range of actions in resonating to weather patterns in areas with a particular micro-climate (Vuojala-Magga et al. 2011). The changes that have taken place in arranging reindeer round-ups naturally affect the structure of *siidas* as well. In Kuttura, it looks as if the older generation is stepping aside as the younger generation demonstrates its success. In the *siida* system, power relationships in governance are not negotiated only by talk but by through actions, and it is the outcome of these actions that guides future decisions.

In the case of carnivore policy, the agent and *siida* system had two ways of reacting historically. First, predators were killed, with the practice nationally encouraged by paying a bounty to hunters. Second, reindeer were herded continuously; small reindeer herders had their reindeer (Inari Sámi) near their homes and large reindeer herders moved around in nomadic fashion with their animals (Fell Sámi). Traditionally, people succeeded in reindeer husbandry by using their individual skills and actions in the struggle with predators. They were successful, as the number of predators was comparatively low from the beginning of the 1900s up to the 1990s.

Today people face demanding situations in protecting their reindeer from predators. In the spring, two families of the Kuttura *siida* tried to herd their reindeer from March until calving time in mid-May. However, in a forested area it is virtually impossible to control the herds, and it is even harder to control the animals during summer time. Individual agents do their jobs as well as they can, yet overall people are helpless. For individual young agents, carrying on reindeer husbandry in this situation is apparently becoming far too risky, and they are losing respect for themselves and their work for the first time in the history of the livelihood. As Öje Danell has said: "The social consequences are devastating due to the traumatic experiences of encountering killed, wounded and disabled animals in the range, strong feelings of insecurity, uneasiness, distrust and powerlessness, as well as amplified conflicts with the surrounding world" (Danell 2009).

Governance in this case is rather anonymous: agents do not know how to act or negotiate, and the individual skills or dynamics of the *siida* are no longer an answer to this problem. There is solidarity between the co-operatives of the South, North, East and West; co-operation between co-operatives has strengthened and the system of balancing gains and losses covers the entire reindeer-herding region. Reindeer herders know that the question of governance is not only a matter of EU or national policy, but that it is also affected by public opinion. Governance in the 2000s has become transformed into a multi-layered system of European media and publicity, which has produced faceless threats and fear. Instead of constructive discussions of indigenous Sámi culture and its valuable and dynamic kinship system based on reindeer husbandry, what one has seen are unsophisticated discussions with one-sided opinions in which predators are victimised and herders are portrayed as criminals. Today reindeer herders have to place more trust than ever in their own organisations, the RHA and BAF, and look to national politics to acknowledge the problem of predators before it is too late.

In Finland, there is a long history of different types of forces governing reindeer husbandry even before the time of EU. World War Two brought various regulations for herders and reindeer husbandry suffered considerable losses. Cyclical weather patterns have caused problems in handling reindeer. However, reindeer herders have always trusted their skills and social systems for their economic survival during times of hardship. Governance as an outside force in reindeer husbandry has been accepted in the same way as changes in weather patterns. Until today reindeer herders have had strong confidence in their actions. Their response to outside forces has been a flexible self-governance system based on the *siida* and its agents, with good skills as means of adaptation. People have been at home in the wilderness, they have had a good knowledge of the behaviour of reindeer and other animals and they have survived in extreme weather conditions. With modernisation, new technology has been adopted in a rather easy manner: snowmobiles have improved the standard of living and the younger generations have trained themselves to use computer programs for bookkeeping and other financial tasks. Development has not hindered the younger generation from taking up reindeer husbandry; the main factor has been their self-confidence as agents, which is based on their belief trust in enskilment and the flexibility of the *siida* system.

Governance on the local level can be seen in the governance of the *siida* system and its dynamics. The power relations among individuals and families produce both tension and competition which motivate individual actions and in this way produce the dynamics inside the *siida*. However, in this dynamic of change and competition, one has to have the skills discussed above in order for the co-operation or competition to function. This produces a self-governing element in *siidas* if observed from the outside. These dynamics are powerful inside the livelihood itself, enabling it to support future generations, co-operatives, with *siidas* serving as strong and dynamic systems that regulate and accept new improvements alongside old approaches as a means of adaptation and transformation.

References

Baer, L. A. (2010). *Study on the impact of climate change adaptation and mitigation measures on reindeer herding*. In Permanent forum on indigenous issues of the United Nations Economic and Social Council. Ninth session. E/C.19/2010/15. http://library.arcticportal.org./686/1/Baer_ LarsAnders_reindeer_climate_change_adaptation.pdf. Retrieved 28 Aug 2011.
Beach, H. & Stammler, F. (2006). Human-Animal relations in pastoralism. In F. Stammler & H. Beach (Eds.), *People and Reindeer in Move. Special Issue of Journal of Nomadic Peoples*. *10*(2), 5–29. Oxford: Berghahn.
Danell, Ö. (2009). Petoeläintilanne Ruotsin poronhoidossa (Predators in Swedish reindeer herding). *Poromies, 5*, 18.
Danell, Ö., & Norberg, H. (2010). Petoeläintilanteen ja liikennevahinkojen vaikutukset Suomen porotalouden teurasmääriin vuosina 2005/06–2008/09 (The effects of the carnivore population and animal deaths in traffic on slaughter quantities in the period 2005/2006–2008/2009; predators and traffic damages). *Poromies, 6*, 15–21.
Eálat home pages (2010). http://www.arcticportal.org/community/ealat. Retrieved 28 Aug 2010.
Eira, I. M. G., Magga, O. H., Mahtis, B. P., Sara, M. N., Mathiesen, S. D., & Oskal, A. (2009). *The challenges of Arctic reindeer herding: The interface between reindeer herders' traditional knowledge and modern understanding of the ecology, economy, sociology and management of Sámi reindeer herding*. EÁLAT project. http//www.arcticportal.org/550/01/Eira_127801.pdf. Retrieved 28 Aug 2011.
Hygieniapaketin asetukset eli Euroopan parlamentin ja neuvoston asetukset EPNA Ey (2004). EU Hygiene regulation. http://europa.eu/legistlation_summaries/food_safety/veterinary_checks_ and_food_hygienie/f84002_fi.htm. Retrieved 28 Aug 2011.

Forbes, B. C., & Stammler, F. (2009). Arctic climate change discourse: The contrasting politics of research agendas in the West and Russia. *Polar Research, 28*, 28–42.

Heikkinen, H. (2002). *Sopeutumisen mallit. Poronhoidon adaptaatio jälkiteolliseen toimintaympäristöön Suomen läntisellä poronhoitoalueella (Models of adaptation to post-industrial society in Finland's western reindeer herding district)*. Helsinki: SKS.

Heikkinen, H. (2006). Neo-Entrepreneurship as an adaptation model of reindeer herding in Finland. *Nomadic peoples, 10*(2), 198–208.

Ingold, T. (2000). *The perception of the environment: Essays in livelihood, dwelling and skill*. New York: Routledge.

Joks, S. (2000). *Tradisjonelle kunskaper i bevegelse: om kontinuiteten i reindriftas praksiser* (Hovedfagsoppgave i sosialantropologi). Tromsø: Univesitetet i Tromsø, Norway.

Juuso, T. A. (2010). *Välikysymys* 16.4.2010 (Comment in the Finnish Sami Parliament by Tuomas Aslak Juuso on Saami reindeer herding). http://www.samediggi.fi/index.php?option=com_docman&Task=cat_view&gid=174&Item=10. Retrieved 28 Aug 2010.

Kainulainen, P. I. (2008). *Maasuurpetojen kielteisten vaikutusten vähentämisen kompensoinnin mahdollisuudet – Esiselvityshanke*. Loppuraportti. Final report on compensation for damage caused by predators. Suomussalmi: Suomussalmen kunta.

Kainulainen, P. (2011). *Selvitys petojen aiheuttamien vahinkojen vaikutuksista porotalouteen ja toimenpiteet pedoista aiheutuvien ongelmien ratkaisemiseksi*. Report on losses to caused by carnivores reindeer husbandry and measures to solve carnivore-related problems. Rovaniemi: Lapin liitto.

Kavakka, M. (2009). Porot, pedot ja uusi riistanhoitolaki (Reindeer, predators and the new game law). *Metsästäjä-lehti, 2*.

Kojola, I. (2009). Suurpedot ja poronhoito – Asian monta puolta (Predators and reindeer herding). *Poromies, 5*, 19.

Kortesalmi, J. (2003). *Poronhoito (Reindeer herding)*. Helsinki/Jyväskylä: SKS.

Kortesalmi, J. (2007). *Poronhoidon synty ja kehitys Suomessa (Reindeer herding and its development in Finland)*. Tampere: Tammer-Paino Oy.

Kozulin, A. (1996). The concept of activity in Soviet psychology. Vygotsky, his disciples and critics. In H. Daniels (Ed.), *Introduction to Vygotsky* (pp. 99–121). London: Routledge.

Lave, J. (1990). The culture of acquisition and the practice of understanding. In J. W. Stiegler, R. A. Shwede, & G. Herd (Eds.), *Cultural psychology: Essays on comparative human development*. New York: Cambridge University Press.

Lapin Liitto (2011). Petofoorumi II 11.10.2011. Kestävällä petopolitiikalla kohti elinvoimaista porotaloutta. (From sustainable predator politics towards vital reindeer management) http://www.lapinliitto.fi/c/document_library/get_file?folderId=223971&name=DLFE-10007.pdf. Retrieved 8 Jan 2012.

Lave, J. (1991). *Cognition in practice: Mind, mathematics and culture in everyday life*. Sydney: Cambridge University Press.

Lave, J., & Wenger, E. (2001). Legitimate peripheral participation in communities of practice. In R. Harrison (Ed.), *Supporting lifelong learning: Perspective on learning* (Vol. 1, pp. 111–126). Florence: Routledge.

Lehtola, V. P. (1997). *Saamelaiset: Historia, Yhteiskunta, Taide (The Sámi: History, society, art)*. Jyväskylä: Gummerus.

Magga, J. A. (2009, April). *Näkkälän paliskunnasta* (Esitelmätiivistelmä. Poropäivät, Kaamanen). Abstract of presentation at the annual reindeer herders' meeting. http://www.rktl.fi/uploads/Seminaarit/esitelmätiivistelmät_poropaivat_2009.pdf. Retrieved 28 Aug 2011.

Magga, J. (2010, April). *Suomen porosaamelaisten puheenvuoro Rovaniemellä 16.4.2010 YK:n erikoislähettiläs James Anayelle* (The statement of Sámi reindeer herders of Finland to UN special ambassador James Anaye, Rovaniemi 16.4.2010). http//:icr.arcticportal/org/index.php?option=com_content&view=article&id=1349%3. Retrieved 5 Jan 2012.

Mazzullo, N. (2010). More than meat on the hoof? Social significance of reindeer among Finnish Saami in a rationalized pastoralist economy. In F. Stammler & H. Takakura (Eds.), *Good to eat, good to live with: Nomads and animals in Northern Eurasia and Africa* (pp. 101–119). Sendai: Sasaki Printing & Publishing.

Meristö, T., Järvinen, J., Kettunen, J., Nieminen, M. (2004). *Porotalouden tulevaisuus – "Mitkä ovat mahdolliset maailmat?" Skenaarioluonnosten esittely* (The future of reindeer herding – opportunities and scenarios). Kala- ja riistaraportteja nro 315. Helsinki: RKTL.

Ministry of Agriculture and Forestry (1996). *Suomen maasuurpetokannat ja niiden hoito* (Predators in Finland and their management). Suurpetotyöryhmän raportti 6/1996. Helsinki: MMM.

Ministry of Agriculture and Forestry (2009). Muistio 26.6.2009. 1718/447/2009. httip://wwwb.mmm.fi/el/laki/lausuntapyynnöt/MMMlammasadressin-taustamuistio.pdf. Retrieved 28 Aug 2011.

Norberg, H. (2010). Maasuurpedot ja porotalouden kannattavuus (Predators and profitability of reindeer herding). *Poromies, 4,* 17–21.

Ollila, T. (2009). Reviiriperusteinen korvausjärjestelmä – kokemuksia kymmenen vuoden jälkeen (Experiences of compensation system – ten years after). *Poromies, 5,* 20–21.

Oskal, N. A. (1995). *Det rette, det gode og renlykken.* Ph.D. thesis, University of Tromsø, Norway.

Paine, R. (1994). *Herds of the tundra: A portrait of Saami reindeer pastoralism.* Washington, DC: Smithsonian.

Pallson, G. (1991). Enskilment at sea. *Man, 29*(4), 901–27.

Raito, K. A., & Heikkinen, H. (2003). Enemmän oma-aloitteisuutta, vähemmän valitusta. Hallinnon näkemyksiä poronhoidon osallistumiseen perustuvien instituutioiden kehittämiseen (More initiative, less complaint. Views of administration on the development of participatory reindeer herding management). In *Technology, society, environment.* Espoo: Otamedia Oy.

Rees, W. G., Stammler, F. M., Danks, F. S., & Vitebsky, P. (2008). Vulnerability of European reindeer husbandry to global change. *Climatic Change.* doi:10.1007/s10584-007-9345-1.

Riista- ja kalatalouden tutkimuslaitos, RKTL (2007). *Suurpedot* (Predators). www.rktl.fi/riista/suurpedot. Retrieved 28 Aug 2011.

Saarni, K., Setälä, J., Aikio, L., Kemppainen, J., & Honkanen, A. (2007). The market of reindeer meat in Finland. Scare resource – high-valued products. *Rangifer Report, 12,* 79–83.

Sara, M. (2001). *Reinen – et gode fra vinden.* Karasjok: Davvi Girji.

Schatzki, T., Knorr-Cetina, K., & von Savigny, E. (2001). *The practice turn in contemporary theory.* London: Routledge.

Tauriainen, J. (2009). Porotalouden kannattavuuskehitys *2002/2003–2008/2009* (Profitability of reindeer herding 2002/2003–2008/2009). *Poromies, 3.*

The Hartwell Paper (2010). A new direction for climate policy after the crash of 2009. http://eprints.lse.ac.uk/27939/1/HarwellPaper_English_version.pdf. Retrieved 28 Aug 2011.

Tuomisto, J. (2008). Sopimustuotannon mahdollisuudet ja rajoitteet poronlihamarkkinoilla (The opportunities and limits of contractual production in reindeer meat markets). In A. Hoppanen (Ed.), Maataloustieteen päivät 10.-11.1.2008. *Suomen maatalousseuran tiedote* 23, pp. 6–12.

Tuomisto, J., & Jauhianen, L. (2008). Poronlihan sopimustuotanto. In L. Rantamäki-Lahtinen (Ed.), *Porotalouden taloudelliset menestystekijät* (pp. 100–109) (The success factors in reindeer herding). MTT:n selvityksiä 156. Helsinki: MTT.

Tyler, N. J. C., Turi, J. M., Sundset, M. A., Bull, K. S., Sara, N. M., Reinert, E., Oskal, N., Nellemann, C., McCarthy, J. J., Mathiesen, S. D., Martello, M. L., Magga, O. H., Hovelsrud, G. K., Hanssen-Bauer, I., Eira, N. I., Eira, I. M. G., & Corell, R. W. (2007). Saami reindeer pastoralism under climate change: Applying a generalized framework for vulnerability studies to a sub-arctic social-ecological system. *Global Environmental Change, 17,* 191–206.

Vuojala-Magga, T., Turunen, M., Ryyppö, T., & Tennberg, M. (2011). Resonance strategies of Sámi reindeer herders in Northernmost Finland during climatically extreme years. *Arctic, 64*(2), 227–241.

Informants

Reindeer herder 1950a male
Reindeer herder 1950b male
Reindeer herder 1948 male

Part IV
Towards a Practice Theory of Adaptation Governance

Chapter 7
Responsibilisation for Adaptation

Heidi Sinevaara-Niskanen and Monica Tennberg

Abstract This chapter concludes the study on adaptation governance for climate change from the perspective of practice theory. The authors suggest that through practices the uncertain future of climate change and its impacts in the Arctic are made "governable" and "governed". Through practices of "distant" governance, reflected in plans and strategies for adaptation, a range of actors become involved in adaptation governance, with varying relationships to the state. A second form of governance, "intimate" governance works by dispersing involvement in adaptation into the communities through multiple, daily interactions. In addition to "distant" and "intimate" governance, there are emerging practices of "close" governance. Through ethical self-creation – a practice of the self – people make themselves subjects and objects of climate change adaptation and its governance. Responsibilisation is a particular technique of power which works by scattering governance and responsibility.

Keywords Practice theory • Adaptation • Responsibilisation • Finland • Russia • Reindeer herding • Floods

7.1 Practices of Adaptation

Why study climate change adaptation at different levels? What does such a multi-level approach tell us about adaptation to climate change impacts? The point of such inquiry is to make sense of adaptive activities as practices at various levels – international, national, regional and local. On the surface, these levels seem quite

H. Sinevaara-Niskanen (✉) • M. Tennberg
University of Lapland, P.O. Box 122, 96101 Rovaniemi, Finland
e-mail: heidi.sinevaara-niskanen@ulapland.fi; monica.tennberg@ulapland.fi

parallel and autonomous, but they are in fact connected by existing not only practices of governance, but also possibly new practices to be developed for adaptation governance. Adaptation has already required, and will continue to require, a rethinking of our daily practices. New approaches will challenge earlier understandings of problems, the division of labour between actors and activities, and the roles and responsibilities of agents. Practices play a central role in adaptation to climate change in that they help us "make sense about the world and about actions" (Schatzki 1996, 111).

People use their present practices and create new ones in order to adapt to changing conditions. Practice theory sees adaptation to climate change as historically and spatially contextualised: adaptation takes place locally but within the frameworks and structures constituted by local, regional and national histories. In addition, practice theory suggests that adaptation governance is taking place in space- and time-specific contexts of governance customs, political cultures and administrative traditions. On balance, practices render the questioning of, as well as governance and agency related to, climate change and its impacts, "governable" and "governed".

First, the practices of the UNFCCC and its national applications construct what the limits of "sayable" and "doable" in adaptation governance are. The problem of climate change and its impacts are defined not only as dangers and threats, but also as "governable" risks. Second, through practices of governance, climate risks are rendered objects of particular kinds of governmental interventions at various levels. Practice theory suggests that the "international" ideas of the IPCC and UNFCCC about adaptation turn into concrete applications and actions at the various levels of governance through emerging practices of adaptation. Finally, the theory also suggests that adaptation governance constructs certain kinds of agencies for the public as objects of adaptive interventions and as subjects of adaptive actions.

For purposes of the present research, the most important aspect of responsibilisation to consider is how practices of questioning, governance and agency construct responsibility in adaptation (Pellizzoni 2004; Summerville et al. 2008; O'Neill et al. 2008; Cruikshank 1999). Practices of responsibilisation render climate change and its impacts a problem for all of us. Questions about what the problem is, how it is to be governed, and by whom all construct responsibilities. Ulrich Beck (1992) claims that it is typical of environmental problems and their management that responsibility for them "disappears" into structures of political governance. As a result, we have a widely spread "organised irresponsibility" in industrialised societies. This can be attributed to the nature of the problems in such societies, for they are not spatially defined or easily calculated, and the solutions do not lie in the hands of individual national states. Climate change as a problem is a problem for everyone and thus for no one. This research project has challenged the notion that the responsibility for adaptation is lost: The studies seek to trace the processes of defining, bearing and sharing responsibility for the governance of adaptation occasioned by climate change and its impacts.

7.2 Practices of Problem-Shaping

Making an issue into a problem or experiencing it as a problematic situation, following Deweyan logic is the starting point for discussing responsibility. At the end of the day, the issue of climate change is a problem of responsibility: Is it a natural or human-induced phenomenon; which activities and which countries are most responsible for it; and who is responsible for taking action to mitigate harmful emissions and advance adaptation to inevitable impacts of climate change in different parts of the world? For the Foucauldian thinker, climate change is a particular governmental problematisation, an apparatus that is made up of many elements and defines multiple responsibilities.

The Finnish and Russian cases of national adaptation governance discussed in the present volume reveal clearly distinct problematisations of adaptation to climate change. In the Russian problematisation, the future of the country's Arctic region is central, yet the question of how vulnerable the Russian Arctic is remains unclear. Defining vulnerability is tantamount to taking responsibility for the region and its future. The Russian Arctic will clearly witness the impacts of climate; however, the number of indigenous people in the region – some two million – is considered "small". The Russian formulation of vulnerability where the Arctic is concerned reflects that the country has yet to decide what to do with the region and does not want to take responsibility for it. The Soviet era produced heavy industrial structures and settlement patterns to support them, and these extended to remote Arctic regions. However, it is uncertain how these remote regions, peoples and livelihoods can be supported and maintained in the future. In the recently published Russian Arctic Strategy, the Arctic is defined as a resource area for the rest of the country (Riabova 2010; Tennberg and Rakkolainen 2010).

In the local Russian case, floods on the Tatta River in Siberia have challenged local norms and practices. People have lost important places – not merely homes, but also sites and environmental features that conveyed a sense of community and identity. Among other natural changes, the floods have been framed as 'disastrous'. They create stress, disrupt normal social processes, and force people to adapt by making temporary adjustments or permanent changes in their everyday practices. Despite their having long-term effects, floods are defined as emergencies in Russia: The threat of flooding is defined not as a risk but as an uncertain danger. This means that people accept some uncertainty and show tolerance for the danger of a flood in terms of the disruption, annoyance and trouble it causes. The problem of flooding has not developed locally into a risk with a market value.

In the national Finnish case, the problematisation of vulnerability – and thus responsibility – is the opposite of that found in Russia. It is a rather recent idea that Finland could actually be vulnerable to impacts of climate change. Now, with the publication of the ACIA (2004), a new governmental problematisation has emerged acknowledging that the people and nature in northern Finland could be vulnerable to those impacts. The focus of concern is reindeer husbandry (and the tourism industry) in Finnish Lapland and their capabilities to adapt to changing environmental conditions.

The irony compared to the Russian case is that the concern in Finland centres on the mere 4,000 Sami reindeer herders in Finnish Lapland.

In the local Finnish case, climate change and its impacts in the form of changing seasons challenge current reindeer herding work and meat production practices. It is the younger generation of reindeer herders in particular who experience the pressures of having to make their livelihood economically viable. Climate change is yet another stress factor to affect many herders, one among numerous other changes taking place in reindeer husbandry. Herders have to negotiate new practices in their *siida*, since without young reindeer herders, the future of the *siida* and the livelihood is dubious. The changes requiring adaptation by reindeer herders and *siidas* are not only external, but also internal, for they challenge reindeer herding practices. The common responsibility of a *siida* – both the younger and older generations – is to ensure the continuity of the livelihood.

These problematisations at various levels – local, regional and national – can be seen both as "a tacit part of common practices of a people and the habits induced by participation in the same observance" (Turner 1994, 106). However, such practices of problematisation also have a compelling power: They seem to entrap the participants in particular ideas of impacts and their significance, thus precluding "multiple interpretations" of changing conditions (ibid., 106). Practices of problem-shaping result in hybrid understandings of climate change adaptation as a problem.

7.3 Practices of Governance

Following Arun Agrawal (2005), practices of governance can be divided into distant and intimate. These practices follow both Deweyan and Foucaultian approaches to governance. Through practices, a "being" is shaped in a thinkable and manageable form for governance. Moreover, as "such beings are formed, ... authorities responsible for them are made (Rose 1999, 22). Multilevel practices of adaptation governance are emergent governmental inventions in global climate governance. These levels should not be taken as given or natural: they are constantly negotiated and reconstructed. From the governance point of view, climate change is a governmental concern and responsibility, a particular milieu made up of responsibilities, rights, and duties. Foucault (2007, 22) refers to Moheau's statement that "[i]t is up to the government to change the air temperature and to improve the climate". Moheau continues: "well, if there has been so much change, it is not the climate that has changed; the political and economic interventions of government have altered the course of things to the point that nature itself has constituted for man ... another milieu ...". Moheau uses the word "nature", but Foucault prefers "milieu" instead.

Through practices of governance "at a distance", a range of actors becomes involved in governance, with varying relationships to the government and to the state (Agrawal 2005, 429). Responsibility in climate governance is constructed at the international level as "common but differentiated", but states understand this responsibility differently in their actions. The Russian governmental response has been one of reluctance, even "irresponsibilisation". International concern over the climate mainly serves foreign policy interests. For the moment, Russian adaptation governance appears to be very centralised, a top-down exercise with intellectual, material and organisational resources that are too limited to allow governance to be properly developed and implemented. Much of the Russian adaptation governance is by nature reactive and corrective, focusing on natural hazards and responses to them; it is not proactive, which would include planning to support and advance sustainable development in different parts of the country. In addition, Russian adaptation governance suffers from weak signals and a lack of knowledge between different governmental levels and bodies. In particular, local communities have not emerged as a specific site of adaptation governance in the Russian context. Although adaptation mostly takes place locally, most of the responsibility for adaptation governance lies at the federal level in the form of risk control and management (Didyk 2010).

In contrast, the Finnish government has become almost overly responsible for climate change: Concern over the climate is embedded in the whole society and is used as an argument in various kinds of debates on development, ones not always directly connected to climate change per se. In terms of national practices, Finnish adaptation governance mostly follows the logic of international climate governance, and in the future it is expected to become even more internationally oriented with the development of the EU's adaptation governance. Responsibilities for adaptation governance are divided between different levels of adaptation, but not in the most organised way. Due to this sectoral approach in adaptation governance, the problem at the regional level is the linkages between sectors and levels of activities. Nevertheless, half of the Finnish population is now covered by some kind of regional or local climate strategy that includes adaptation to climate change. Numerous sites – municipalities, regions and organisations – have become governmentalised in the name of concern over the climate. Local and regional adaptation governance follows the national logic: top-down, expert-based and sectoral administrative planning for the future.

Responsibility is no longer understood only as a relationship with the state, but as one of obligation towards those for whom the individual cares most: his or her family, neighbourhood, workplace and, ultimately, community. Accordingly, practices of governance are also "intimate" (Agrawal 2005, 195). Intimate governance shapes practices and helps to knit together individuals in communities as well as their leaders, state officials, and politicians. Intimate governance works "by dispersing rule and scattering involvement in government". It is based on "everyday practices"; that is, "the joint production of interests is based on multiple, daily interactions" (ibid., 195). This suggests a more Deweyan, or deliberative, approach to governance.

A case in point is that the Russian logic of governance does not recognise the link between the construction of dams and the occurrence of floods. The ownership of dams is an open question. Whose property and responsibility are they? Who has economic responsibility for them? It is telling that all water resources in Russia are state property and river embankments are a federal responsibility. Regions participate in water construction projects. Dam safety is a duty of the owner of the dam, who operates the dam, but safety control is the responsibility of the state. In addition, federal ministries participate in the monitoring of water resources. The distant governance of natural disaster emergencies, development projects and water management have diverse practices based on unclear responsibilities.

At the regional level, this same practice of distant governance seems to have resulted in a lack of co-ordination in flood prevention measures in the Tatta region. Responsibility for the development of water management is also placed on the municipality. The municipal administration, as an essential actor in the community's social network, also acts as a buffering institution between the community's perceptions, ideas and deeds. However, it has limited opportunities for practical flood prevention. Municipal actions are not co-ordinated, which leads to frustration and conflicts when floods actually occur.

The local inhabitants in Tatta deal with the flood threat and actual emergencies through individual and collective practices. In their case, the external and internal pressures seem to create social cohesion in the form of extended social networks and co-operation. The local natural and human resources and the regional ethnic identity form the basis of adaptation. The local people bear the responsibility on their own in the sense that they may accept some everyday risk of flooding and related problems. The practices of intimate governance follow local perceptions. The interviews reveal a profound ambivalence about what to believe and a reluctance to act directly. At the level of intimate governance, the focal issue is to define "good" and "bad" dams in terms of flood prevention, not whether or not to build dams in the first place. This focus redirects the ultimate responsibility for flood prevention and water management to the federal government.

In the local Finnish case, while the *siida* functions as part of the official system of reindeer herding governance in Finland, as a historical form of Sami self-governance it is also external to the system. Nevertheless, it is part of the Finnish distant governance through interactions among the reindeer herding co-operatives, the reindeer herding association and national administration for reindeer husbandry. The *siida* is governed from a distance through EU regulations, funding for meat processing and hygiene practices. Despite its role in distant governance, the *siida* is also a form of "intimate" governance. It is based on a division of labour between skilled reindeer herders and other actors in the *siida* and in the reindeer herding cooperative. The *siida* is embedded in multiple practices of reindeer work, kinship and Sami governance. It is governed by its own practices, and relies on trustful relations between its members.

However, the practices of the *siida* can become challenged by external and internal stresses which while they may sometimes lead to conflicts, may also spawn new forms of collaboration between relatives and individual members. The responsibilities formed by distant and intimate governance converge in the *siida*, which becomes a pressure cooker of sorts. The tension between old and new practices of reindeer herding is based on a kind of natural economy in which "some habits cannot be acquired without displacing others" (Turner 1994, 106).

7.4 Practices of Agency

In addition to distant and intimate governance, there are emerging practices of "close" governance. Through ethical self-creation and practices of the self, people make themselves subjects and objects of climate change and adaptation. Ultimately, adaptation governance produces power and knowledge and environmental subjects whose thoughts and actions bear some reference to climate change (Agrawal 2005, 14). Much of adaptation governance constructs, defines and organises the conduct of subjects and aims at mobilising them to become the active, knowledgeable and responsible citizens required for the practice of the participative "democracy" that is needed for sustainable development (Summerville et al. 2008, 697, 710).

In "close" governance, "a practice is a set of considerations that governs how people act" and how they govern themselves (Schatzki 1996, 96). Practices are essential for the ways people act and define their agency. Practices rule action "not by specifying particular actions to perform, but by offering matters to be taken account of when acting and choosing" (ibid., 96). The actions taken vary according to the situation and individual interpretations of it, surroundings, and the motives, or "sentiments", behind the selection of actions that are intended to create the conditions sought. Hence, practices "provide considerations" which can be taken into account – or ignored – when choosing the way to act on each particular occasion (Ibid., 96).

What defines practices – including practices of adaptation – is that they are inherently social. Participation in practices "entails entering a complex state of coexistence with other participants" (Schatzki 1996, 169). Hence, practices are also socially negotiated. As Schatzki (1996, 172) describes the process, sociality itself is founded on practices: the "individuals and relations involved exist only within practices". It is also through the sociality of practices that social formations such as families, governments, and economies are constituted; it is thus within practices that social worlds and identities are defined.

The practice of responsibilisation is an integral part of "close" governance: It enables subjects to do their own self-governing in ways that, while according them a degree of autonomy, nonetheless integrate them into a web of power/knowledge and reporting that systematically holds them to self-account (Dillon 1995). Foucault emphasised self-assertion through self-creation, whereas Dewey stressed that self-creation fostered better community and better communion with individuals (Garrison 1998, 112).

This responsibilisation is effected in the name of freedom: "Technologies of agency seek to enhance or deploy individuals' capacities as agents, transforming them into active citizens capable of managing their own risk" (Higgins 2001, 303). According to O'Malley (1996, 199–200), one result is that we have a responsible (moral) and rational (calculating) individual. At the same time the national individual will wish to become responsible for him- or herself by taking steps to avoid and to insure against risk in order to independent. Responsibilisation results in an everyday practice of the self: "Prevention and risk management now become the responsibility" of the individual (O'Malley 1996, 202).

Our theoretical understanding leads us to expect that the prudent subject of risk must be responsible, knowledgeable and rational. However, viewed from this particular perspective in the light of the analysis of national reports and documents, our subjects in the Russian and Finnish cases seem quite "ecologically ignorant" or "indifferent to climate concern". They are knowledgeable about climate change, yet somewhat sceptical about it. Most importantly, they avoid taking responsibility for the actions to tackle climate change that would be expected of "rational" and "prudent" actors.

In general, the Russian population is aware of climate change, but not very concerned about it. Where taking action is concerned, other everyday problems and daily survival dominate the Russian popular views about climate change. People's historical experiences of living under the conditions of the totalitarian system in the Soviet Union, as well as the turbulence of the post-socialist period, define Russian citizenship and its relation to the state. On the one hand, people still believe that the state is obligated to help them and they rely on its actions; on the other, they do not trust the state to be able to provide them with stability. There exists a duality of trust/non-trust, which may lead to inaction towards state authorities. In the case of Tatta, the increasing governmental rhetoric of self-responsibility and shortcomings of state programmes in the management of floods propelled the question of self-reliance and self-governing of risks onto the community's agenda. The questions of how individuals assess and 'process' risk and how they rank risks brings us closer to the multiple meanings associated with risk that underpin people's everyday lives and facilitate their adaptive agency. Adaptive strategies are also the result of a process of innovation through which people build up not only their skills, but also the self-confidence necessary to shape their environment. The building of Tatta regional identity through culture and education has contributed to the building of a responsible community that is prepared to invest in itself and shape strategies for addressing experienced and expected risks.

In the Finnish case, one can see the development of responsible and risk-conscious climate citizens: There is a high level of awareness and willingness to contribute to mitigating the effects of climate change. Also worthy of note is people's willingness to be governed and to be part of adaptation governance. However, Finns in general feel that they lack the means to take action. The allowed scope of citizen action is mostly economic: Individuals are expected to make wise choices in their everyday life based on an exercise of economic reasoning. The traditional scope of political action and participation seems to be limited to official representation by various NGOs in the national and regional structures of decision-making.

In the local Finnish case, the dominating value of the reindeer herding culture is continuity. The *siida*, as a social network, aims at securing the continuity of the livelihood at any cost. However, the reindeer herding culture is also dominated by insecurity, with reindeer herders meeting challenges in their work daily. In this context of continuity/uncertainty, climate change and its impacts are nothing new. Every member of the *siida* bears responsibility for action on his or her own, but within that framework. The *siida* is based on a division of labour between its members: everybody has a task and everybody is needed. The networks of power/knowledge within the *siida* system are based on skills, kinship, and number of reindeer; it is these elements which define individual agencies and responsibilities. Through negotiations in the *siida*, reindeer herders adapt to changes in a "minute-by-minute" mode, trusting in their own skills and the all-embracing support of the community. In some cases, securing continuity requires altruism by the members of a *siida* and even a willingness to accept unpleasant decisions.

The differences in agencies in the Finnish and Russian cases can be understood as differences in "the ways in which the subject is constituted in the power/knowledge networks of culture" (Oksala 2002, 225). Cultures are bound to normalising categories, concepts, images and practices. Being a subject means being subjected by external power, but it is not reducible to that alone. An important part of subjectivity is effected through "internalising the power, through the subject's own conscience and self knowledge" (ibid., 225). It is also important to note that, in Foucault's view, technologies of self do not introduce a totally autonomous subject. This in turn suggests that practices of self-governance are not invented by individuals themselves: "They are patterns that he finds in his culture and which are proposed, suggested and imposed on him by his culture, his society and his social group" (Foucault 1988, 11).

In our cases, Finnish and Russian people are both risk bearers and survivors in changing climatic conditions. A double logic obtains among the Arctic peoples where the practice of responsibilisation is concerned: "[T]he responsibility for the environment is shifted onto the populations[;] citizens are called to take up the mantle of saving the environment in attractively simplistic ways. This allows for the management, self-surveillance and regulation of behaviour in such a way that lays claim to subjectivity…" (Rutherford 2007, 299). Arctic indigenous peoples are called upon as examples and advocates of sustainable development.

7.5 Scattered Responsibilisation as a Practice

Climate change is often described as "a wicked problem" (a concept from planning theory, see Rittel and Webber 1973/1984), one that contests rationalities and practices of governance. It is difficult to define the problem, find solutions for it and define responsibilities for the actions required, as Deweyan approach to problematic situation suggests. Adaptation to impacts of climate change takes place through multiple rationalities. Foucauldian logic suggests that climate change adaptation and governance are always controversial, conflictive, and incomplete. This is true for our case studies. Adaptation governance in Finland and Russia can be described as "scattered". As Agrawal (2005, 195) notes, governance works by "dispersing rule and scattering involvement in government", but it is also rooted in everyday practices and the joint production of interests in multiple daily interactions. Governance practices shape and share the responsibility for climate change adaptation. Responsibility is uneven, fragmented and shifting as a result of continuous negotiations between various social actors and multiple levels of government. The scattering of responsibilities can be likened to building on quicksand.

The problematisation of climate change entails multiple perceptions, observations, and conceptions which are open to different interpretations and practices: Adaptation to climate change can be seen as a scattered problem. Views of climate change challenge the nature of adaptation as a scattered problem, contesting any assertions that it is only external to social groups and structures. The importance of adaptation is not perceived or experienced equally among individuals, livelihoods and communities, nor are its implications for agency self-evident. As adaptation governance develops, the involvement of various societal actors in solving and managing the problem becomes scattered. As a result, the responsibilities for addressing climate change-related problems are scattered as well. Individual agents become scattered agents for adaptation; they are required to become self-reliant and self-responsible and to be able to negotiate their everyday practices in the light of adaptation.

Building on quicksand is hazardous, and might be even impossible. Developing new practices and responsibilities takes time. Agency is a matter of making iterative changes to particular practices. It is about the possibilities and responsibility entailed in refiguring material-discursive practices. Practices are also based on re-education and re-culturation. Characteristic of practices is that they are not "all acquired at once, or in the same place, but there is considerable redundancy in the 'lessons' one learns about power" (Turner 1994, 113). What is learned is also individual. However, once learned, practices can not be easily unlearned. "The reason the process seems hidden is precisely that there is no magic moment at which the lessons are learned, and no single point at which they are 'transmitted'" (ibid., 113). This describes very well the process of adaptation to climate change and the related practices. It is very difficult to pinpoint a particular event or moment when adaptation takes place and new practices are formed.

The two perspectives in the present research on climate change adaptation and governance, the Deweyan and Foucauldian, illuminate a specific, non-essentialist approach to studying how people govern themselves. A major connection that can be made between Deweyan pragmatism and Foucauldian governmentality is a common refusal to abide by the rules of essentialist discourses. Both Dewey and Foucault suggest that ethico-political practice must be conceived of as the cornerstone of experience. This practice encompasses both theoretical knowledge and an embodied sense of living in the corporeal world. As with Foucault, the self in Dewey's ethics is not a substance, but a form, or an organisation of habits that are relatively enduring but subject to change (Reynolds 2004).

Both thinkers also suggest that the potential for resistance is neither exclusively individual nor exclusively collective, but rather that these two dimensions constitute the critical potential of subjectivities. For Foucault, resistance lies in the ethical self-creation grounded in a multiplicity of individual everyday practices, while for Dewey individuals cannot be divorced from their social surroundings. Resistance can be found in scattered problematisations, as well as in involvement in government adaptation planning and various everyday actions. For example, the superficial indifference to climate change and its impacts seen among the local populations in the Arctic, at least in our case studies, can be interpreted as resistance in different ways. On the one hand, it can be seen as a fatalist world-view in which climate change is not an anthropogenic problem and is thus outside the sphere of human action. On the other hand, it can be interpreted as resistance to a globally produced and constructed environmental concern. Then again, the view can be interpreted as a strong belief in individuals' capacities to tackle future challenges. Resistances, too, are often built on quicksand.

Our case studies reveal that the rationality of climate change adaptation relies on multiple and scattered responsibilisations. In the case of Russia, responsibilisation is unclear and undefined – it can be called even "irresponsibilisation" – whereas in Finland the process is very much top-down and, as a result, society is "almost overly responsibilised" at least in terms of governmental practices. Responsibility, however, is re-negotiated in various sites in government; it is bound to space, place and time. Even when the challenges are the same for communities, their responses may differ considerably. The political traditions, cultures and practices in our case studies show the bounded rationalities of adaptation governance. Responsibilities are inscribed into practices slowly. "The natural economy of practices", made up of political, administrative, social and cultural traditions of governance, determines the adoption of new practices. Responsibilisation works as a particular technique of power by scattering governance and responsibility.

References

ACIA (2004). Arctic climate impact assessment. http://amap.no/acia/. Retrieved 17 May 2010.
Agrawal, A. (2005). *Environmentality. Technologies of government and the making of subjects*. Durham/London: Duke University Press.
Beck, U. (1992). *Risk society*. London: Sage.
Cruikshank, B. (1999). *The will to empower: Democratic citizens and other subjects*. Ithaca: Cornell University Press.
Didyk, V. (2010, December). Formation of local self-government and administrative reform in Russia: Aims and reality – case of the Murmansk region. Presentation at the BIPE workshop, Rovaniemi, Finland.
Dillon, M. (1995). Sovereignty and governmentality: From the problematics of new world order to the ethical problematics of the world order. *Alternatives, 20*, 323–368.
Foucault, M. (1988). The ethic of care for the self as a practice of freedom (interview). In J. Bernauer & D. Rasmussen (Eds.), *The final Foucault*. Cambridge, MA: MIT Press.
Foucault, M. (2007). *Security, territory and population* (Lectures at the Collège de France 1977–1978). Houndmills/New York: Palgrave-McMillan.
Garrison, J. (1998). Foucault, Dewey and self-creation. *Educational Philosophy and Theory, 30*(2), 111–134.
Higgins, V. (2001). Calculating climate: Advanced liberalism and the governing of risk in Australian drought management. *Journal of Sociology, 37*(3), 299–316.
O'Malley, P. (1996). Risk and responsibility. In A. Barry, T. Osborne, & N. Rose (Eds.), *Foucault and political reason: Liberalism, neo-liberalism and rationalities of government* (pp. 189–208). London: University College London Press.
O'Neill, J., Holland, A., & Light, A. (2008). *Environmental values*. London/New York: Routledge.
Oksala, J. (2002). *Freedom in the philosophy of Michel Foucault*. Helsinki: University of Helsinki.
Pellizzoni, L. (2004). Responsibility and environmental governance. *Environmental Politics, 13*(3), 541–565.
Reynolds, J. M. (2004). "Pragmatic humanism" in Foucault's later work. *Canadian Journal of Political Science, 37*(4), 951–977.

Riabova, L. (2010, December). *State policy in the Russian North, its social outcomes and needs for change.* Presentation at the BIPE workshop, Rovaniemi, Finland.

Rittel, H., & Webber; M. (1973/1984). Dilemmas in a general theory of planning. Reprinted in N. Cross (Ed.), *Developments in design methodology* (pp. 135–144). Chichester: Wiley. http://www.uctc.net/mwebber/Rittel+Webber+Dilemmas+General_Theory_of_Planning.pdf. Retrieved 17 June 2011.

Rose, N. (1999). *The powers of freedom.* Cambridge: Cambridge University Press.

Rutherford, S. (2007). Green governmentality: Insights and opportunities in the study of nature's rule. *Progress in Human Geography, 31*(3), 291–307.

Schatzki, T. R. (1996). *Social practices: A Wittgensteinian approach to human activity and the social.* Cambridge: Cambridge University Press.

Summerville, J., Adkins, B. A., & Kendall, G. (2008). Community participation, rights and responsibilities: The governmentality of sustainable development policy in Australia. *Environment and Planning C: Government and Policy, 26,* 696–711.

Tennberg, M., & Rakkolainen, M. (2010). Venäjä ja ilmastonmuutos. *Idäntutkimus, 4,* 50–54.

Turner, S. (1994). *The social theory of practice: Tradition, tacit knowledge and presuppositions.* Cambridge: Polity Press.

Index

A
Adaptation governance, 5–8, 10–13, 18–23, 25, 26, 29–31, 40, 42–50, 89–97, 104, 125–129, 131–135
Adaptive capacity, 4, 41, 49, 117
Adaptive co-management, 18, 25
Agency, 6–10, 13, 25–30, 43, 58, 72–74, 77–80, 88, 95, 102, 116–120, 126, 130–134
Agro-pastoralism, 56, 62, 66, 73
Alaas meadows, 60–62, 76
  ethnic identity, 62
  products quality, 60–62
  phenology, 76
Arctic, 3–5, 7–12, 17, 19, 22, 39–42, 45, 46, 48, 50, 59, 64, 88, 89, 96, 114, 115, 125, 127, 133, 134

C
Carnivore, 102, 112–116, 119
Climate change, 3–13, 17–30, 39–50, 58, 64, 86–97, 101–120, 125–135
Climate citizenship, 26, 95, 96, 132
Community, 4, 6–8, 10–12, 20, 22, 23, 25–27, 29, 30, 41, 42, 46, 49, 50, 55–80, 88, 91, 92, 96, 104, 117, 127, 129–132, 134, 135
  culturally 'affiliated', 62–63
  marginalised, 63, 70
  knowledgeable, 60, 72–73, 77, 80

D
Dam, 42, 64, 70, 78, 79, 129

Development projects, 70, 77, 78, 129
Dewey, J., 6, 12, 20, 22–24, 26, 27, 29, 30, 131, 134
Disaster, 47, 55, 58, 59, 66, 67, 69, 71, 73, 74, 76–78, 129
  definition, 65–66
  'irrational' return, 74–77
  solidarity, 73–74

E
Ecological citizenship, 26
Emergency, 6, 12, 19, 21, 40, 43, 44, 47, 59, 66–74, 127, 129, 130
European Union (EU), 4, 9, 13, 46, 86, 88, 89, 91, 92, 97, 101–120, 129, 130

F
Federal government, 70, 130
Finland, 9, 10, 12, 85–91, 94–97, 104–106, 111, 113, 115, 119, 127, 128, 130, 133, 135
Flooding, 55–80
Foucault, M., 12, 21, 22, 24, 25, 27–30, 128, 131, 133, 134

G
Gibson (theory of affordances), 59, 77
Governance, 5–13, 17–31, 39–50, 58, 68, 85–97, 102, 104–106, 116–120, 126–135
Governmentality, 6, 23–25, 28, 30, 134

I
Identity, 58, 62, 80, 127, 130, 132
Irrigation, 56, 64, 77, 78

K
Kuttura, 10, 102, 103, 105, 111–112, 114, 115, 118, 119

L
Lapland, 7, 10, 12, 13, 86, 88, 89, 91–95, 97, 113, 127, 128
Local government, 7, 47, 68

M
Murmansk, 12, 40, 45–48, 50

P
Path-dependency, 18
Perception, 58, 59, 64–66, 71, 77, 78, 93, 130, 133
Planned adaptation, 5, 10, 22, 42
Practices, 4–13, 18, 19, 21–23, 25, 27–31, 40, 56, 59, 60, 68, 69, 73, 76, 79, 86, 97, 102, 103, 106–112, 125–135
Practice theory, 13, 19, 29, 126
Problem, 3, 7, 9, 10, 12, 19–26, 29, 30, 40–44, 49, 76, 77, 86–89, 93–95, 102, 112–116, 119, 126–129, 133, 134

R
Resilience, 19, 101, 117
Regional government, 68, 69
Reindeer herding, 7, 10, 89, 92, 97, 101–120, 128, 130, 132
Reindeer herding collective, 10
Responsibilisation, 13, 19, 28, 95–97, 125–135
Responsibility, 5–7, 11, 13, 17–19, 21–26, 29, 30, 50, 68–71, 78, 79, 95, 96, 126–135
Resistance, 28, 134–135
Risk, 4, 7, 12, 18, 21, 27, 43, 48, 49, 56, 58, 66–71, 74, 77, 79, 87, 126, 127, 129–133
Rovaniemi, 86, 91, 92, 97, 108

Rule-based theory, 129
Russia, 5, 9, 10, 12, 39–50, 56, 59, 61, 64, 66, 69, 77, 127, 129, 133, 135

S
Sakha, 10, 56, 59, 60, 62, 63, 66, 68, 69, 71, 73
Siida, 10, 13, 102–120, 128, 130, 132, 133
Self-governance, 10, 28, 30, 101, 104, 117, 120, 130, 133
Solidarity, 73, 74, 119
State, 5–7, 9, 12, 20, 22–26, 28, 30, 31, 43, 47, 49, 50, 59, 60, 63, 64, 66–73, 78–80, 128, 129, 131, 132

T
Tatta River, 10, 12, 56, 57, 64, 68, 72, 77, 79, 127
Tourism, 42, 45–47, 86–88, 92–93, 96, 127
Trust and distrust, 70–73

U
Uncertainty, 4, 20, 28, 58, 72, 73, 79, 87, 89, 127, 132
UNFCCC. *See* United Nations Framework Convention on Climate Change
United Nations Framework Convention on Climate Change (UNFCCC), 4, 5, 24, 25, 86, 87, 89, 126
Uolba, 10, 59, 61–65, 72

V
Vulnerability, 8, 10, 19, 41, 87, 88, 127

W
Water management, 10, 56, 70, 77–79, 129, 130

Y
Ytyk Kyöl, 10, 56, 57, 59, 63–65, 68–74, 78, 79